Economics and Ecology

ECONOMICS AND ECOLOGY

New frontiers and sustainable development

Edited by

Edward B. Barbier

Senior Lecturer
Department of Environmental Economics
and Environmental Management
University of York
UK

CHAPMAN & HALL
London · Glasgow · New York · Tokyo · Melbourne · Madras

Published by Chapman & Hall, 2–6 Boundary Row, London SE1 8HN

Chapman & Hall, 2–6 Boundary Row, London SE1 8HN, UK

Blackie Academic & Professional, Wester Cleddens Road, Bishopbriggs, Glasgow G64 2NZ, UK

Chapman & Hall Inc., 29 West 35th Street, New York NY10001, USA

Chapman & Hall Japan, Thomson Publishing Japan, Hirakawacho Nemoto Building, 6F, 1–7–11 Hirakawa-cho, Chiyoda-ku, Tokyo 102, Japan

Chapman & Hall Australia, Thomas Nelson Australia, 102 Dodds Street, South Melbourne, Victoria 3205, Australia

Chapman & Hall India, R. Seshadri, 32 Second Main Road, CIT East, Madras 600 035. India

First edition 1993

© 1993 Chapman & Hall

Typeset in 10/12 pt Palatino by ROM-Data Corporation, Falmouth, Cornwall

Printed in Great Britain by T J Press, Padstow

ISBN 0 412 48180 4

A catalogue record for this book is available from the British Library

Library of Congress Cataloging-in-Publication data available

∞ Printed on acid-free text paper, manufactured in accordance with the proposed ANSI/NISO Z 39.48–1992 and ANSI Z 39.48–1984 (Permanence of Paper)

Contents

Contents

Contributors

Edward B. Barbier
Senior Lecturer,
Department of Environmental
Economics and Environmental
Management,
University of York
UK

Scott Barrett
Assistant Professor of Economics,
London Business School
UK

Stephen M.J. Bass
Associate Director,
Forestry and Land Use Programme,
International Institute for Environment and Development

Gardner Brown Jr
Professor of Economics,
University of Washington
USA

Gordon R. Conway
Vice Chancellor of the University of Sussex,
Sussex House, University of Sussex
UK

Robert Costanza
Director,
Maryland International Institute for Ecological Economics
UK

Anil Markandya
Harvard Institute for International Development
London
UK

D.W. Pearce
Director,
Centre for Social and Economic Research on the Global Environment
(CSERGE),
University College London and University of East Anglia
UK

Charles Perrings
Professor,
Department of Environmental Economics and Environmental Management,
University of York
UK

Ian Scoones
Research Associate,
London Environmental Economics Centre
UK

Ian R. Swingland
Founder and Research Director,
The Durrell Institute of Conservation and Ecology
UK

R.K. Turner
Executive Director,
Centre for Social and Economic Research on the Global Environment
(CSERGE),
University of East Anglia
UK

Preface

In the Summer of 1991, Bob Carling, who was then Life Sciences Editor of Chapman & Hall, approached me over the possibility of producing an edited volume of works on economics and ecology. As we discussed the matter further, what became clear is that there is a growing literature on the 'frontiers' of both disciplines that has involved economists 'borrowing' from ecology and ecologists 'borrowing' from economics. We decided that this volume should try to provide a small cross-section of that literature.

I was very much interested in editing this volume for several reasons. First, one of my principal interests in economics has been how the economic analysis of natural resource and environmental problems can benefit from the concepts and lessons learned from other disciplines, in particular ecology. I was grateful at having the opportunity to pull together a selection of readings that illustrate how the integration of the two disciplines can lead to fruitful analysis. Second, I was also aware that, as Director of the London Environmental Economics Centre and as a Senior Researcher at the International Institute of Environment and Development, I was fortunate to have worked with or to have known a number of economists and ecologists whose work would be ideal for this volume. I was delighted that so many of my friends and colleagues were as enthusiastic about this project as I, and agreed to participate. Finally, I felt that there would indeed be interest in such a volume from economists, ecologists and others because of recent attempts to establish an intellectual 'dialogue' between the two disciplines. Of the fora that I have participated in, the two that stand out most in my mind are the International Society of Ecological Economics, and their new journal *Ecological Economics*, and the Beijer International Institute of Ecological Economics of the Royal Swedish Academy of Sciences. Hopefully, there will be more in the future.

One cannot, of course, ignore recent developments on the policy-making front. As I am writing this, the June 1992 'Earth Summit' at Rio – the UN Conference on Environment and Development – has just concluded. At the core of its major charter of principles, *Agenda 21*, the UNCED has firmly identified the need to integrate the environment into all levels of economic decision-making as the basis for 'operationalizing' sustainable development. To do this, will clearly require economics, ecology and other disciplines being applied in a systematic and coherent way to real-world development problems. If we are to understand how best to tackle environment and development issues, then it is important that we first begin to understand each other. My hope is that economists and ecologists will both find this book of interest, and more importantly, that they begin to explore what insights the other discipline can offer to their own.

EB
London

Acknowledgements

Since its inception in October 1988, the London Environmental Economics Centre (LEEC) has been at the forefront of the application of economic analysis to environment and development problems. Under LEEC's auspices, much work has been done involving economics and ecology. The majority of authors in this volume are either members of LEEC – directly employed or as Associate Fellows – or have collaborated with LEEC in the past in one form or another. It is safe to say that, without LEEC, this volume may not have been put together. I am therefore deeply indebted to the Swedish International Development Authority, the Netherlands Ministry of Foreign Affairs and the Norwegian Ministry of Foreign Affairs, who have shown consistent faith in LEEC by providing core funding.

LEEC is an integral part of the International Institute for Environment and Development (IIED). IIED has provided the multi-disciplinary setting within which much of LEEC's work on environmental economics has thrived – including the lessons learned from ecology. I am pleased that both current and former members of IIED have contributed to this volume. Ultimately, however, it is the whole Institute and its Executive Director, Richard Sandbrook, whom I must thank for their unwavering support of LEEC and their willingness to employ economics in a multi-disciplinary context.

Bob Carling of Chapman & Hall was the prime mover behind this volume. He encouraged and nurtured it from the beginning – let us hope his faith has been justified. Vinetta Holness and Jacqueline Saunders of LEEC used their skills ably in helping me make this volume presentable. I am grateful to their assistance. Although they did not contribute directly to this volume, Bruce Aylward, Jo Burgess, Josh Bishop and Michael Collins of LEEC have been a source of inspiration on many research ideas in recent years.

Finally, I would like to thank the contributors themselves, whom I also consider my friends and colleagues. In urging them to complete their chapters, I hope I have not used up too much of my 'capital' with them. I hope the others do not mind if I single out David Pearce and Anil Markandya, with whom collaboration in economics has been so rewarding, and Gordon Conway, who has shared so much of his knowledge of ecology with me in the past.

EB

1

Introduction: economics and ecology – the next frontier

Edward B. Barbier

1.1 A SHORT PARABLE

At meetings where I am speaking to a 'mixed' audience of economists, ecologists and researchers from other disciplines, I sometimes like to begin my talk with the following tale:

One day, an economist was walking across a huge parking lot. Out of the corner of her eye, she spied a ten dollar bill. She stooped down to pick it up, but then stopped herself. Scratching her head, she muttered to herself, 'Oh no, the probability of my finding a ten dollar bill in this parking lot is too small!' As the economist walked off, she was soon followed by an ecologist. He saw the ten dollar bill and also stopped to pick it up. But in turning it over, he noticed some condensation, a little moss and some tiny organisms underneath the bill. 'My goodness!', he exclaimed. 'There is a whole ecosystem under here!' The ecologist soon became lost in observation so that he failed to notice the ten dollar bill blowing away. The bill soon blew across the path of an environmentalist. He saw the ten dollars and picked it up. 'Tsk, tsk!', he proclaimed to himself. 'The amount of litter in this car park is shameful! ' And so he threw the ten dollar bill away in the nearest waste basket.

I find the above parable a useful reminder of how economists, ecologists and environmentalists can all look upon the same object or situation – e.g. finding a ten dollar bill in a parking lot – but view it from a completely different perspective. And, of course, all three end up responding differently,

Economics and Ecology: New frontiers and sustainable development.
Edited by Edward B. Barbier. Published in 1993 by Chapman & Hall, 2–6 Boundary Row, London SE1 8HN. ISBN 0 412 48180 4.

not only compared to each other but also to what the 'ordinary person in the street' would do.

The purpose of the following chapters is not to emphasize the differences in perspectives between economies and ecology, but to show how each discipline can in fact learn from the other's perspective. Despite the diversity of topics covered in this volume, all the chapters essentially revolve around one overriding theme – how can we best 'operationalize' sustainable development – which requires a careful appreciation of economic, environmental and social parameters. It is not surprising, therefore, that the authors – whether economists or ecologists – are all attempting to incorporate these parameters in their analysis and work. As a consequence, the resulting chapters are a testimony as to how economists are beginning to appreciate and employ ecological principles in their thinking of environment and development, and equally, how ecologists are beginning to adopt economic 'thinking' to bolster their perspective on similar problems.

Recent developments in the economic concept of 'natural capital' and its application to environment and development problems of the countries of the South is a good example of the convergence between economic and ecological thinking.

1.2 NATURAL CAPITAL, ECONOMICS AND ECOLOGY IN DEVELOPMENT

Economists generally consider an economy's 'endowment' of natural resources and ecosystems as a potential source of 'wealth'. This is particularly apparent in the developing regions of the world, where much economic production and livelihoods are dependent on natural resource exploitation. Like other 'assets' in developing economies, environmental resources can be seen as a form of 'natural' capital. That is, they have the potential to contribute to the long-run economic productivity and welfare of developing countries. Thus the value of a natural resource as an economic asset depends on the present value of its income, or welfare, potential.

However, in any growing economy there will be other assets, or forms of wealth, that yield income. Any decision to 'hang on' to natural capital therefore implies an opportunity cost in terms of forgoing the chance to invest in alternative income-yielding assets, such as man-made 'reproducible' capital. If natural resources are to be an 'efficient' form of holding on to wealth, then they must yield a rate of return that is comparable to or greater than that of other forms of wealth. In other words, an 'optimal' strategy for a developing country would be to 'draw down' its stock of natural capital to finance economic development by reinvesting the proceeds in other assets that are expected to yield a higher economic return. Under such circumstances, natural resource depletion is economically justified; it should proceed up to the point where the comparative returns to 'holding on' to the remaining natural capital stock equal the returns to alternative investments in the economy. If the latter always exceeds the former, then in the long run even complete depletion of natural capital is economically 'optimal'. The key economic issue is whether the process of 'drawing down' natural capital to accumulate other economic assets is done in the most efficient manner possible.

The idea that economic well-being, or welfare, may not be affected and may even be enhanced if the rents derived from depleting natural capital, such as tropical forests, wetlands, biodiversity, energy, minerals and even soils, are reinvested in reproducible capital has been around for some time in the theoretical economics literature. For example, what is now known as the 'Hartwick–Solow rule' states that, if the rents derived from the inter-temporally efficient use of exhaustible natural resources are reinvested in reproducible – and hence non-exhaustible – capital, it is possible to secure a constant stream of consumption over time (Solow, 1974 and 1986; Hartwick, 1977). Similarly, basic renewable resource economic theory suggests that, for slow growing resources such as tropical forests, it may under certain conditions be more economically optimal to harvest the resource as quickly and efficiently as possible and reinvest the rents in other assets that increase much faster in value. Equally, if the harvesting costs are low, or the value of a harvested unit is high, then the resource may also not be worth holding on to today (Clark, 1976; Smith, 1977).

However, there are obvious limits to the applications of the above rules to real world environment and development problems. For example, the Hartwick–Solow rule assumes that there is sufficient substitutability between reproducible (man-made) and natural capital over time such that they effectively comprise a single 'homogeneous' stock. Moreover, the above rules assume that all economic values are known and reflected in the 'prices' of resources, markets are undistorted, resource extraction is efficient and rents are reinvested in other assets in the economy. As a consequence, more recent economic theories now stress the limits to 'substitution' between many forms of natural and man-made capital, even for developing countries interested in 'drawing down' their natural capital stock in favour of investing in other forms of capital (See Chapter 2 by Barbier and Markandya, as well as Barbier, 1989; Bojö, Mäler and Unema, 1990; Pearce, Barbier and Markandya, 1990). From an ecological standpoint, a major factor is the failure to consider the economic consequences of the loss in 'resilience' – the ability of ecosystems to cope with random shocks and prolonged stress – that results from natural capital depletion (see, in particular, Chapter 4 by Conway, as well as Common and Perrings, 1992; Conway, 1985; Conway and Barbier, 1990; Holling, 1973).

For example, we are now learning from ecology and other disciplines that certain functions of a tropical forest, such as its role in maintaining micro-climates, protecting watersheds, providing unique habitats and supporting economic livelihoods of indigenous peoples, may be irretrievably lost when the forest is degraded or converted. Often the economic values of these and other functions of a tropical forest are not properly accounted for in decisions concerning forest use. There is uncertainty over many of these values, such as the forest's role in maintaining biodiversity and global climate (see the discussion in Chapter 7 by Swingland). Finally, there is little evidence to suggest that tropical forest resources are currently being exploited efficiently, nor that the rents earned from activities that degrade or convert the forests are being reinvested in more 'profitable' activities. Rather, market and policy failures are often rife, and economic rents tend to be dissipated and misused.[1]

In addition, the degradation of resources in many low income countries is often occurring in 'fragile' environments – semi-arid rangelands, 'marginal' cropland, coastal ecosystems, converted forest soils. If resource degradation affects the 'resilience' of these environments to respond to further stresses (e.g. soil erosion, devegetation, etc.) and random shocks (e.g. drought, pest attacks), then the economic livelihoods of peoples dependent on these environments can be affected. If carrying capacities and ecological tolerance limits are transgressed, then collapse of these economic –environmental systems is a possibility (see, in particular, Chapters 5 and 6, by Perrings and Scoones, respectively).

In sum, although it is theoretically possible that the current rates of resource depletion in developing regions are economically 'optimal', it is unlikely that this is the case. Moreover, both economists and ecologists are urging caution in assuming that current rates and natural resource degradation and depletion are either efficient or sustainable. Too often, decisions concerning natural capital depletion and conversion are taken without considering the opportunity cost of these decisions. A major reason for this 'policy failure' is the inadequate application of ecological and economic analysis to determine the full range of economic benefits provided by natural resources and ecosystems.[2]

In short, we often do not know whether it is worth 'holding on' to the natural capital as an economic asset because we do not know or bother to take into account the potential economic benefits that it yields. As a result,

---

[1] For further discussion of these issues in relation to tropical forests, see Barbier (1992c).

[2] Yet there are a growing number of examples of economic analysis – in conjunction with empirical ecological studies – being applied in developing countries to assess the environmental costs and benefits of investment and planning decisions. This has occurred at the level of **national accounts** (e.g., see Ahmad, El Serafy and Lutz 1989; Repetto *et al.*, 1989; TSC/WRI, 1991), **investment projects** (Dixon *et al.*, 1988) or **individual resource systems** – such as tropical wetlands (Barbier, Adams and Kimmage, 1991; Barbier, 1993 Ruitenbeek, 1991), tropical forests (Barbier, 1992b; Peters, Gentry and Mendelsohn, 1989; Ruitenbeek, 1989), drylands (Bojö, 1991; Dixon, James and Sherman, 1989 and 1990), wildlands (Dixon and Sherman, 1990 and Swanson and Barbier, 1992), agroforestry (Anderson, 1987 and Barbier, 1992a) and watersheds (Easter, Dixon and Hufschmidt, 1986; Gregersen *et al.*, 1987).

decisions will always be biased towards environmental degradation because the underlying assumption is that the forgone benefits provided by natural capital in developing countries are necessarily negligible. If the opposite is the case, as much economic and ecological evidence suggests, then current levels of resource depletion in developing countries may actually be 'excessive' – a problem that stems not just from the inefficient transformation of natural capital into other forms of capital but also from its 'unsustainable' management for economic welfare.

Economics, ecology and the role of natural resources in 'sustaining' economic development are all underlying themes of this volume of papers. Moreover, what we would like to show is not the disparity of views on this theme, but the convergence of economic and ecological principles towards similar conclusions.

1.3 OUTLINE OF THE BOOK

Most of the chapters in this volume address the role of natural resources in 'sustaining' development, by focusing on applications in developing countries. This is not surprising, given the pervasive role of natural resource management issues in many development problems encountered in the South. Moreover, it is in tackling the complex issues posed by environment and development that much 'cross-fertilization' between economics and ecology has occurred. We hope that the issues raised by these chapters and the analytical results will be of interest to researchers seeking to apply economics and ecology to other natural resource management problems than the ones discussed in this volume.

Chapter 2 by Edward Barbier and Anil Markandya continues the theme of 'natural capital and development' raised in this chapter by developing an analytical framework of the relationship between environment and sustainable economic activity. A key feature of the analysis is the assumption that natural resources and processes are essential to human welfare. Thus the 'sustainability' of the economic–environment system is dependent on whether certain **biophysical constraints** are observed by the economic process. The authors then examine the different conditions that determine whether or not these constraints are observed, and the system remains 'sustainable' over the long term. The theoretical insights provided by this model are an important introduction to the remaining chapters in the volume, many of which provide specific examples of the type of economic–environmental systems and resource management problems that are susceptible to these conditions.

In Chapter 3, Robert Costanza further explores recent developments in our conceptions as to how economic and ecological systems work. The chapter describes some of the emerging tools available for systems analysis, and their limitations in the face of scientific risk and uncertainty. By drawing on a number of applications, Costanza argues that the failure of policymakers to take into account adequately the limited predictability of complex economic–environmental systems may lead to the choice of incorrect policy recommendations and conclusions.

Systems analysis – more precisely agroecosystem analysis – and its applications to agriculture in developing countries is also the topic of Chapter 4 by Gordon Conway. For many years, the author has been at the forefront of the application of agroecosystem analysis to determine the sustainability of farming systems, whether at the field, household or regional level. Conway adapts the ecological concept of 'resilience' for his definition of sustainability, which he suggests is the ability of an agroecosystem to withstand disturbing forces – particularly threats to its overall productivity. These may consist of both **shocks** (e.g. infrequent floods or droughts, new pest outbreaks or a sudden rise in input prices) and **stresses** (e.g. soil erosion, increased salinity, declining market demand). The social value of an agroecosystem can then be determined by analysing how its inherent sustainability is traded off against other important properties, such as productivity, equity and stability.

Charles Perrings provides an economic interpretation of the concept of 'sustainability as resilience' in agroecosystems in Chapter 5. Perrings develops a bioeconomic feedback control model of rangeland management in semi-arid areas – a management problem characterized by a high level of uncertainty associated with high variance in rainfall and the interdependence of the dynamics of herd size and carrying capacity. The model is used to analyse the long term effects of different price structures for livestock offtake on the size of herd and carrying capacity of the range, given the ecological uncertainty that results from stresses and shocks on the system. The analysis reveals the extent to which the economic–environmental system is sensitive to key economic variables, emphasizing the need for policymakers to take into account this effect in their development of rangeland policies.

In devising his model, Perrings discusses both ecological and economic carrying capacity in pastoral systems. In Chapter 6, Ian Scoones focuses on these two concepts in detail, employing the findings of field work carried out in communal areas in Zvishavane District, Zimbabwe. The focus is on a mixed agropastoral system of dryland cropping and livestock production – particularly the role of the latter as inputs into agriculture. The main purpose of the chapter is to clarify confusion over the various concepts of livestock carrying capacity, which in the past have resulted in inappropriate measures of productivity and policies. For example, whereas **ecological carrying capacity** is determined by environmental factors, i.e. the maximum number of animals the land can hold without being subject to density dependent mortality and irreversible environmental degradation, **economic carrying capacity** is determined by economic factors, i.e. the stocking rate that offers maximum economic returns to producers, given their economic objectives and definition of 'productivity'. Scoones examines both concepts of carrying capacity in order to determine whether communal area farmers' economic strategies are ecologically sustainable, and whether existing analyses based on these concepts are assessing accurately the ecological impacts of these strategies.

In Chapter 7, Ian Swingland addresses another issue of major topical importance – the role of tropical forests in biodiversity conservation. Current definitions of biodiversity are rather ambiguous. Swingland argues that a more 'scientifically sensible' meaning is required if we are to understand the link between biodiversity and tropical forests and develop appropriate policies. He reviews what we know about this linkage, and the key policy and planning issues that have arisen. Although based on an ecological perspective, the chapter also discusses the role that empirical economics research can contribute to biodiversity and tropical forest conservation.

Chapter 8 by Scott Barrett complements the preceding chapter by modelling explicitly the economic–ecological relationship between biodiversity, the optimal stock of forest area conserved (i.e. 'wildlife' preserves) and deforestation. Biodiversity is determined by an ecological species – area relationship, which is in turn affected by the amount of forest land that is converted to alternative use (e.g. timber production and agriculture). Consumption, and thus welfare, depends on both production from converted forest land (e.g. agriculture) and the rate of deforestation (e.g. timber harvesting), but the stock of biodiversity in the remaining forest area also influences welfare. Thus the relationship between 'development' benefits and 'preservation' benefits determines the optimal provision of forest reserves and hence biodiversity conservation. This problem is analysed by Barrett for the case where underlying benefits and costs of preservation are independent of time, and where there is technical progress.

A major difficulty in determining 'optimal' wildlife preserves and protected areas is that the economic benefits associated with environmental conservation are often 'non-marketed'. This equally applies to the 'enjoyment' experienced by tourists from visiting protected areas and viewing wildlife. Individuals appear to get more enjoyment, or value, from this experience than what they usually pay for it in terms of park fees and expenditures on travel and lodgings. The 'extra' value gained over costs is an estimate of the net economic benefits, or 'consumer surplus', derived from wildlife viewing. In Chapter 9, Gardner Brown, with the assistance of Wes Henry, estimates the value of viewing elephants in Kenya through the application of travel cost and contingent valuation methods. By using these methods, the total net economic value per foreign visitor on a wildlife viewing safari is calculated, and a portion of that total value is then allocated to viewing elephants specifically. The chapter also discusses the impacts of wildlife safari expenditures on tourism revenues and employment in Kenya, and the likely implications of declining elephant populations due to poaching.

Chapter 10 by Steve Bass discusses what many believe to be an ideal 'research station' for observing ecological and economic interdependencies – the small island economic–environmental system. Small islands comprise mutually interdependent economic, social/demographic, cultural, political, physical and ecological subsystems. Disequilibrium – or 'unsustainability' – in the overall island economic–environmental system is prevalent in many cases, and results when one of its component subsystems (e.g. ecology or economy) replaces another too rapidly, with inadequate time for all the subsystems to adjust. This is often the result of the high exposure of island ecologies, economies and societies to external influences, and the low capacities for adjustment in relatively small, resource-poor islands. Bass explores this theme with reference to the commodity booms and collapses of island economies and the consequent development of the 'MIRAB' (migration, remittances, aid and bureaucracy) economy of the Caribbean.

Finally, Chapter 11 by Kerry Turner and David Pearce returns to the concept of sustainable development as a means to exploring the interface between economics, ecology and ethics. The conventional economic paradigm of utilitarian cost–benefit analysis is modified to allow for intergenerational equity, through incorporation of a 'constant natural capital' rule. The modified paradigm, which the authors call the 'sustainability paradigm', can be adapted for several conditions varying from **very strong sustainability** to **very weak sustainability**. The paradigm is capable of capturing comprehensively the anthropocentric values associated with the environment, and has the potential for capturing at least part of the intrinsic values identified by bioethicists. The authors conclude that, from an ecological and economic standpoint, the 'sustainability paradigm' can in most practical instances offer sufficient safeguards to the environment that further consideration of the bioethical standpoint is not necessary.

In sum, the following chapters offer much evidence of the practical benefits that can result from economists and ecologists learning from each other's perspectives. Perhaps if policymakers begin listening to the results that such research is producing, we will advance further towards the goal endorsed by the recent UNCED 'Earth Summit' of 'integrating environment in all levels of decision-making'.

REFERENCES

Ahmad, Y., El Serafy, S. and Lutz, E. (1989) *Environmental Accounting for Sustainable*

Development, A UNEP–World Bank Symposium, The World Bank, Washington DC.

Anderson, D. (1987) *The Economics of Afforestation,* Johns Hopkins University Press, Baltimore.

Barbier, E.B. (1989) *Economics, Natural-Resource Scarcity and Development: Conventional and Alternative Views,* Earthscan, London.

Barbier, E.B. (1992a) Rehabilitating gum arabic systems in Sudan: Economic and environmental implications, *Environmental and Resource Economics,* **2,** 19–36.

Barbier, E.B. (1992b) Tropical deforestation, in *Economics for the Wilds: Wildlife, Wildlands, Diversity and Development* (eds T. Swanson and E.B. Barbier), Earthscan, London.

Barbier, E.B. (1993) Sustainable use of wetlands – valuing tropical wetland benefits: economic methodologies and applications. *The Geographer's Journal,* **159,** 22–32.

Barbier, E.B., Adams, W.M. and Kimmage, K. (1991) *Economic Valuation of Wetland Benefits: The Hadejia-Jama'are Floodplain, Nigeria,* LEEC Discussion Paper 91–02, London Environmental Economics Centre, London.

Bojö, J. (1991) *The Economics of Land Degradation: Theory and Applications to Lesotho,* The Economic Research Institute, Stockholm School of Economics, Stockholm.

Bojö, J., Mäler, K-G and Unemo, L. (1990) *Environment and Development: An Economic Approach,* Kluwer Academic, Dordrecht.

Clark, C. (1976) *Mathematical Bioeconomics: The Optimal Management of Renewable Resources,* John Wiley, New York.

Common, M. and Perrings, C. (1992) Towards an ecological economics of sustainability, Draft paper, Centre for Resource and Environmental Studies, Australia National University, Canberra.

Conway, G.R. (1985) Agroecosystem analysis, *Agricultural Administration,* **20,** 31–55.

Conway, G.R. and Barbier, E.B. (1990) *After the Green Revolution: Sustainable Agriculture for Development,* Earthscan, London.

Dixon, J.A., Carpenter, R.A., Fallon, L.A., Sherman, P.B. and Manipomoke, S. (1988) *Economic Analysis of the Environmental Impacts of Development Projects,* Earthscan, London.

Dixon, J.A., James, D. and Sherman, P. (1989a) *The Economics of Dryland Management,* Earthscan, London.

Dixon, J.A., James, D. and Sherman, P. (eds) (1989b) *Dryland Management: Economic Case Studies,* Earthscan, London.

Dixon, J.A. and Sherman, P. (1990) *Economics of Protected Areas: Benefits and Costs,* Island Press and Earthscan, London.

Easter, K.W., Dixon, J.A. and Hufschmidt, M.M. (1986) *Watershed Resources Management: An Integrated Framework with Studies from Asia and the Pacific,* Westview Press, Boulder, Colorado.

Gregersen, H.M., Brooks, K.N., Dixon, J.A. and Hamilton, L.S. (1987) *Guidelines for Economic Appraisal of Watershed Management Projects,* FAO, Rome.

Hartwick, J.M. (1977) Intergenerational equity and the investing of rents from exhaustible resources, *American Economic Review,* **66,** 972–4.

Holling, C.S. (1973) Resilience and the stability of ecological systems, *Annual Review of Ecological Systems,* **4,** 1–24.

Pearce, D.W., Barbier, E.B. and Markandya, A. (1990) *Sustainable Development: Economics and Environment in the Third World,* Edward Elgar, London.

Peters, C., Gentry, A. and Mendelsohn, R. (1989) Valuation of an Amazonian rainforest, *Nature,* **339,** 655–6.

Repetto, R., Magrath, W., Wells, M., Beer, C. and Rossini, F. (1989) *Wasting Assets: Natural Resources in the National Income Accounts,* World Resources Institute, Washington DC.

Ruitenbeek, H.J. (1989) *Social Cost–Benefit Analysis of the Korup Project, Cameroon,* Prepared for the World Wide Fund for Nature and the Republic of Cameroon, London.

Ruitenbeek, H.J. (1991) *Mangrove Management: An Economic Analysis of Management Options with a Focus on Bintuni Bay, Irian Jaya,* Prepared for EMDI/KLH, Jakarta, Indonesia.

Smith, V.L. (1977) Control theory applied to natural and environmental resources: An exposition, *Journal of Environmental Economics and Management,* **4,** 1–24.

Solow, R.M. (1974) Intergenerational equity and exhaustible resources, *Review of Economic Studies,* Symposium on the Economics of Exhaustible Resources, 29–46.

Solow, R.M. (1986), On the intertemporal allocation of natural resources, *Scandinavian Journal of Economics,* **88** (1), 141–9.

Swanson, T.M. and Barbier, E.B, (1992) *Economics for the Wilds: Wildlife, Wildlands, Diversity and Development,* Earthscan, London.

TSC/WRI (1991) *Accounts Overdue: Natural Resource Depreciation in Costa Rica,* World Resources Institute, Washington DC.

2

Environmentally sustainable development: optimal economic conditions[1]

Edward B. Barbier and Anil Markandya

2.1 INTRODUCTION

A review of the literature on 'sustainable' economic development suggests that two interpretations of that concept have emerged: a wider concept concerned with sustainable economic, ecological and social development and a more narrowly defined concept largely concerned with 'environmentally sustainable development', i.e. with optimal resource and environmental management over time.[2]

This chapter is concerned with the 'narrower' interpretation – the relationship between environmental quality and sustainable economic activity. The latter is associated with the policy objective of maximizing the net benefits of economic development, subject to maintaining the services and

[1] Earlier versions of the model presented in this paper appeared in Barbier (1989a) and Barbier and Markandya (1990).

[2] For example, the wider, highly normative view of sustainable development was endorsed by the World Commission on Environment and Development (WCED, 1987), which defined the concept as 'development that meets the needs of the present without compromising the ability of future generations to meet their own needs'. The recent UN Conference on Environment and Development also takes this view, as does the Second World Conservation Strategy (UNCED, 1992 and IUCN/UNEP/WWF, 1991). For an economic interpretation of this wider view see Barbier (1987) and chapter 11. For further discussion of the more 'narrow' interpretation of environmentally sustainable development, see Barbier (1990a); Common and Perrings (1992); Klaassen and Opschoor (1991); Pearce, Barbier and Markandya (1990); Pearce, Markandya and Barbier (1989); and Pezzey (1989).

Economics and Ecology: New frontiers and sustainable development.
Edited by Edward B. Barbier. Published in 1993 by Chapman & Hall, 2–6 Boundary Row, London SE1 8HN. ISBN 0 412 48180 4.

quality of natural resources over time. The term 'natural resources' is used broadly. It includes renewable resources, such as water, terrestrial and aquatic biomass; non-renewable resources, such as land in general, minerals, metals and fossil fuels; and semi-renewable resources, such as soil quality, the assimilative capacity of the environment and ecological life support systems.

As discussed in Chapter 1, it is convenient to think of these resources as collectively comprising the **natural capital stock** of the economy. That is, like other economic assets, they have the potential to contribute to economic productivity and welfare – and this contribution is directly related to the services and quality of the asset over time. However, maintaining the services and quality of natural capital over time is obviously not a costless activity. Investments in this activity are only 'worth it' to an economy if the returns, measured in terms of overall welfare gains, exceed the returns from alternative investments in the economy. The latter are clearly the 'opportunity' costs of 'holding on' to natural capital. That is, the economic resources tied up in maintaining the services and quality of the environment could earn a return, in terms of increased economic welfare, if invested elsewhere. Thus, from a purely economic perspective, the above 'sustainability' condition imposed on economies of 'maintaining the services and quality of natural resources over time' makes sense only if the economic value of natural capital exceeds the opportunity cost of holding onto it.[3] Otherwise, society would be better off 'drawing down' its natural capital and reinvesting resources in other economic investments.

On the other hand, as the arguments in Chapter 1 suggest, one should be cautious in assuming that an irreversible loss of the natural capital stock is justified if it results in capital formation elsewhere in the economy. Some of the functions of the environment are not replicable by reproducible – or man-made – capital, such as complex life support systems, biological diversity, aesthetic functions, micro-climatic conditions and so forth. Others might be substituted but not without unacceptable cost. For many ecological functions and resources, we may not comprehend sufficiently their value in support of economic activities and livelihoods until they are irrevocably lost. In addition, degradation of one or more parts of a resource system beyond some threshold level may lead to a breakdown in the integrity of the whole system, dramatically affecting recovery rates and resilience of the system. The total costs of the system breakdown may exceed the value of the activity causing the initial degradation. The uncertainty

---

[3] Editor's note – the emphasis on economic perspective is extremely important. This is in recognition that the services and quality of natural resources are being judged to have value in terms of human welfare alone; that is, a very narrow **instrumental value** concept of the 'worth' of natural resources is being employed here. However, as Chapter 11 discusses, other concepts of value other than the narrow economic interpretation of value are also applied to the environment – with of course different implications for the management of natural resources.

over the true value of natural capital to present and future generations, the implications of irreversible resource degradation and depletion for human welfare, and the lack of man-made substitutes for many of the economic functions of natural capital suggest a high value should be placed on maintaining the services and quality of natural capital.

Thus under certain conditions maximizing the net benefits of economic development, subject to maintaining the services and quality of the stock of natural resources over time, is an essential criterion for sustainable development. In other words, to prevent economic welfare from declining over time requires keeping the overall level of **environmental quality** intact. This of course assumes that:

1. the services, or functions, of the environment are essential to the economic system;
2. there are insufficient substitution possibilities between reproducible capital and these environmental functions;
3. these environmental functions are not augmented by a constant positive rate of technical progress.[4]

If these conditions hold and maintaining environmental quality is an essential criterion, then certain biophysical constraints need to be observed. That is, if the resource base is a composite of exhaustibles and renewables (including semi-renewables and waste-assimilative capacity), sustainability requires:

1. utilizing renewable resources at rates less than or equal to the natural or managed rates of regeneration;
2. generating wastes at rates less than or equal to the rates at which they can be absorbed by the assimilative capacity of the environment;
3. optimizing the efficiency with which exhaustible resources are used, which is determined, *inter alia*, by the rate at which renewable resources can be substituted for exhaustibles and by technological progress.[5]

---

[4] These conditions follow from an extension of the analysis presented by Dasgupta and Heal (1979) of what constitutes an 'essential' economic input. For example, if the three important economic functions of the environment are broadly considered to be the production of useful material and energy inputs, E_1, the assimilation of waste, E_2, and the provision of human, ecological and life-supporting services, E_3, then each of these E_i functions is essential if in its absence feasible consumption must necessarily decline to zero in the long run. On the other hand, if reproducible capital, or a labour-capital composite can be sufficiently substituted for each E_i, then it is no longer essential. Equally, even in the absence of such substitution possibilities, if each E_i could be augmented by a constant positive rate of technical progress, the loss of that function could be managed to ensure that its technically enhanced services are bounded away from zero. Note also that, if perfect substitution of reproducible for natural capital is not possible, the policy of investing resource rents in reproducible capital as suggested by Hartwick (1977) and Solow (1986) may no longer generate a constant consumption stream.

[5] For more on these biophysical constraints as conditions for sustainability, see in particular Pearce (1987 and 1988). For other discussions of the conditions determining the need for biophysical constraints see Barbier (1989a and 1990a); Common and Perrings (1992); Daly (1987); Page (1977), and Solow (1986).

Failure to obey these constraints will lead to a process of environmental degradation as the resource base is depleted, wastes accumulate and natural ecological processes are impaired.

In this chapter we examine explicitly how a hypothetical economy in which natural capital is essential might respond to the limits imposed by the above biophysical constraints. From this analysis the factors determining optimal sustainable economic growth can be derived. In the next section, a simple model is developed to characterize the conditions necessary to maintain the environmental sustainability of an economic system over time.

2.2 A MODEL OF ENVIRONMENTALLY SUSTAINABLE ECONOMIC ACTIVITY

The following model will be used to analyse optimal growth paths for an economy faced with the choice of operating under the three long-term biophysical constraints: harvesting of renewable resources within their natural and managed rates of regeneration; extracting exhaustible resources at the rate at which renewables can be substituted for them (which over the long run implies a zero rate of exhaustion of the 'composite' resource); and emitting wastes within the assimilative capacity of the environment.

The key to the following model is a particular definition of environmental degradation. At any time t, the rate of degradation S is a function of: (i) the flow of waste (W) in excess of the amount assimilated by the environment (A) and (ii) the flow of renewable resources harvested from the environment (R) in excess of the (managed or natural) biological productivity of these resources (G), plus the flow of exhaustible resources extracted from the environment (E). Mathematically, this may be written as:

$$\dot{S} = f([W - A], [(R - G) + E]) . \tag{1}$$

The following assumptions are made about Equation (1):

1. It is a differentiable, increasing function of its arguments. As the net waste level increases and as the excess rate of harvesting increases, so does the level of environmental degradation.
2. It is a convex function of its arguments. This can best be illustrated in Figures 2.1–2.3. As net wastes emitted increase, so the rate of degradation increases at an increasing rate (Figure 2.1). The same applies to net resources harvested (Figure 2.2). However, there is also a trade-off between these two factors that influence the environment. To attain a constant level of degradation, one can reduce the net harvesting of resources if one increases the net levels of waste generated. But the reduction in harvested resources required as one increases the net waste generated itself increases (Figure 2.3).

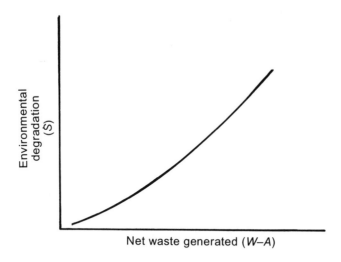

Figure 2.1 dS/dt and *(W–A)*.

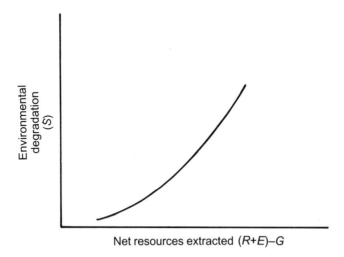

Figure 2.2 dS/dt and $(R+E)–G$.

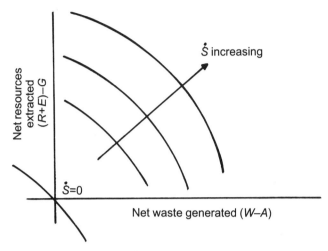

Figure 2.3 Contour curves for dS/dt.

3. The combinations of [W - A] and [(R - G) + E] that achieve zero degrada-
    tion are given by the locus of points going towards the origin in Figure
    2.3. Thus a sufficient condition for zero degradation is $W = A$ and ($R + E$)
    = G.[6]

Note that, as stated, Equation (1) is 'symmetrical'; that is, if the biophysical constraints are observed (i.e. $W \leq A$ and ($R + E$) $\leq G$) in the long run),
then the rate of environmental degradation will be zero or there may be an
'improvement' in environmental quality.[7] These effects will be made more
explicit in the next section. However, it is worth noting that observed
environmental impacts are more likely to be 'asymmetrical'; i.e. in some
economic–environmental systems, it may take a long time before adherence
to these biophysical constraints leads to any improvement in environmental quality, whereas failure to observe these constraints may cause
rapid environmental degradation.

Having defined the level of environmental degradation as a function of

---

[6] As W is total waste from the economic process and thus comprises waste from exhaustible
resource extraction, from renewable resource harvesting and production and consumption, it
may appear that Equation (1) is double counting. But Equation (1) is not accounting for the
flow of material through the economic system but for the total impact on the environment, i.e.
the 'composite' resource base, of the waste generation and resource depletion created by the
economic system. For example, a forest might be depleted faster than its rate of regeneration,
whereas the total waste generated by the harvesting plus production and consumption of the
wood products may not exceed the assimilative capacity of the environment. Looking at the
total impact on the environment of this economic activity therefore requires accounting for
both the impact of the total waste generation – which is negligible in this example – and the
impact of the harvesting – which is significant.

net waste generation and net resources consumed, we now wish to link it to the level of economic activity (consumption) and the 'stock' of environmental assets, or natural capital, available. We label each of these respectively as C and X, and postulate the following functions with the following properties.[8]

$$W = W (C), \ W' (C) > 0, \ W'' (C) > 0 \tag{2}$$

$$R = R (C), \ R' (C) > 0, \ R'' (C) > 0 \tag{3}$$

$$E = E (C), \ E' (C) > 0, \ E''(C) > 0 \tag{4}$$

$$A = A (X), \ A' (X) > 0, \ A'' (X) < 0 \tag{5}$$

$$G = G (X), \ G' (X) > 0, \ G'' (X) < 0 \tag{6}$$

In other words W, R and E are increasing convex functions of C, and A and G are increasing concave functions of X. The concept of environmental quality adopted by this model is fairly broad and essentially synonymous with the entire stock of environmental goods, or natural capital. The three basic functions, or 'services', of this stock are the assimilation of waste, the production of material and energy inputs for the economic system and the provision of amenity, life support and general 'ecological' services. This

---

[7] Note also that the condition $R+E \leq G$ allows for steady substitution of renewables for exhaustible resources as stocks of the latter allow and increases in relative scarcity occur. Thus, as argued by Pearce (1987), if the resource base is viewed as a composite of renewables and renewables, if users are indifferent between which is used, and if the renewable resource use rate should never exceed the regeneration rate, then exhaustibles can safely be diminished by current generations. As a result, the composite stock of resources can be maintained across generations, even though in physical terms current extractions of non-renewables will reduce the stock available to future generations. Moreover, as argued by Hartwick (1977) and Solow (1986), if this composite stock also includes any reproducible capital, and assuming reproducible and natural capital are fully substitutable, then a policy of investing rents from natural resource depletion in reproducible capital can also imply that the composite stock is being maintained intact. However, the assumption in our model is that there are no substitution possibilities between reproducible and natural capital.

[8] Following Becker (1982) and Mäler (1974) it is assumed that environmental quality – our X variable – is measured by a stock of environmental goods that yield a flow of services proportional to that stock in each time period. However, Becker defines this stock variable as 'the differences between the level of pollution for which life ceases and the current level of pollution'. Similarly, Mäler considers only the quality and flow of waste residuals and recycling to have an impact on environmental quality in his intertemporal models. Here it is assumed that environmental quality may be affected not only by net waste generation but also by net resource depletion, as both of these may contribute to environmental degradation if biophysical limits are exceeded (see Equation (1)). This implies a fairly broad, but perhaps more realistic, concept of the 'stock' of environmental assets. For a given type of ecosystem with its associated energy flow, a measure of environmental quality may include, in addition to Becker's definition, the ecosystem's biomass, i.e. the volume or weight of total living material found above or below ground, plus some measure of the distribution of nutrients and other materials between the biotic (living) and abiotic (non-living) components of the ecosystem.

allows the model to assume that, for all intent and purposes, A and G are increasing functions of X. The assumption that W, R and E are all increasing functions of economic activity are well known results stemming from material-balance models.[9] In addition, the assumption of the convexity of waste generation, $W(C)$, and the concavity of assimilative capacity, $A(X)$, is consistent with the models developed by Forster (1973 and 1975), which examine optimal economic growth under one set of these biophysical constraints, namely that waste levels should not exceed the assimilative capacity of the environment. Similarly, standard models of renewable resource harvesting assume concave growth functions, $G(X)$, and convex harvesting rates, $R(C)$.[10]

Substituting (2)–(6) into (1) yields an equation in which dS/dt is a function of C and X. Since A and G are concave functions, it follows that $-A$ and $-G$ are convex. In addition we wish to impose a minimum value, $\underline{X}$, which we define as the **minimum environmental stock** that provides a viable base for sustained economic activity. Hence dS/dt, which is an increasing convex function of its original arguments, $[W-A]$ and $[(R-G) + E]$, will also be a convex function in its transformed state, as a function of C and X, for X greater than $\underline{X}$. This can be written as

$$\dot{S} = h(C, X), \text{ for } X \geq \underline{X}, \tag{7}$$

$$\dot{S} \gg 0, \qquad \text{for } X < \underline{X}.$$

where $h_c > 0$, $h_{cc} > 0$, $h_x < 0$ and $h_{xx} > 0$ and $h_{cc}h_{xx} - h_{xc}^2 > 0$.

Finally, a link can be established between $\dot{S}$, the rate of environmental degradation, and $\dot{X}$, the rate at which environmental quality is changing. Clearly, the basic relationship is an inverse one: as the degradation increases so the environmental stock declines. Assuming both are measured in comparable units, we can represent this by

$$\dot{X} = -aS \tag{8}$$

$$\dot{X} = -ah(C, X), \qquad \text{for } X \geq \underline{X} \tag{9}$$

$$\dot{X} \ll 0, \qquad\qquad \text{for } X < \underline{X}$$

where a is a constant scalar. A typical set of contours of (9) are given in Figure 2.4. As consumption increases, so the environmental quality adjusted stock required to keep the level of degradation constant increases at an increasing rate.

The dynamics of these relationships can best be illustrated in the case of

[9] See, for example, d'Arge and Kogiku (1973); Kneese, Ayres and d'Arge (1970); Mäler (1974); and Victor (1972).
[10] See, for example, Clark (1976), Dasgupta (1982) and Smith (1977).

agriculture – or perhaps more appropriately agroecosystems – as one is talking about systems that are directly dependent on environmental resources and essential ecological functions for 'sustainability'. Thus, for example, in Chapter 4, Conway employs the ecological concept of 'resilience' as the basis of his definition of agricultural sustainability – 'the ability of a system to maintain its productivity when subject to stress or shock'. Unchecked resource abuse within an agroecosystem, whether the result of inappropriate use of agro-chemicals and fertilizers, overcropping of erodible soils, poor drainage, etc., can affect overall agroecosystem sustainability by increasing the susceptibility to stress, shock, or both. The key is reducing the resource degradation, and therefore the stresses and shocks associated with it, to a level where the natural processes and functions of the agroecosystem – appropriately subsidized by human-made inputs and innovations – can counteract these disturbances and thus preserve overall sustainability. Perrings and Scoones, in Chapters 5 and 6 respectively, examine different aspects of stress, shock and resilience in analysing pastoral systems in semi-arid zones of sub-Saharan Africa.

Barbier (1990b) has also applied a relationship similar to (9) in a model of soil erosion in the uplands of Java. Here, the environmental stock

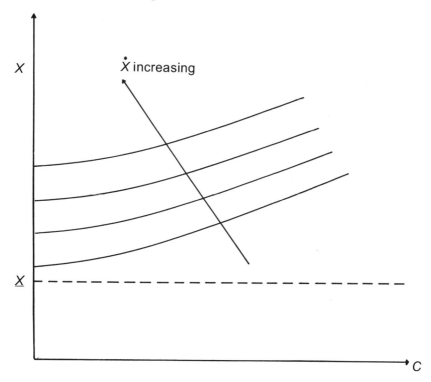

Figure 2.4 C,X loci for given dX/dt.

variable, X, is a measure of soil depth, which is degraded at an accelerating rate by the use of a conventional cropping system and is augmented by the use of an alternative soil conservation package. In this system, the cross-partial derivatives between these two arguments of the function h were considered to be negative. Barrett, in Chapter 8, employs a similar – albeit more explicit – relationship to (9) in his model of optimal conservation of biological diversity. In his model, species diversity is an exponential function of habitat area, which in turn is decreased through the rate of land development. Finally, there is the possibility that the strict conditions on an economy imposed by biophysical constraints may hold for an 'isolated' economic – environment system. Chapter 10 by Bass gives every indication that the ecology of small island economies may be particularly susceptible to the type of environmental degradation relationship indicated by (9).

2.3 OPTIMAL SUSTAINABLE ECONOMIC GROWTH

Using the above model, it is now possible to explore the conditions for environmentally sustainable growth. For example, it is assumed that social welfare at any point in time is measured by a strictly concave utility function U of current C and the current stock of X:

$$U = U(C, X), \tag{10}$$

with $U_c > 0$, $U_{cc} < 0$, $U_x > 0$, $U_{xx} < 0$, $U_{cx} = 0$, $\lim_{c \to 0} U_c = \infty$, and $\lim_{x \to 0} U_x = \infty$.

Equations (1) and (7) were deliberately constructed to reflect the sustainability criteria of observing the biophysical constraints. That is, a minimum condition for an economic growth path to be sustainable over the long run is $W = A$, $R + E = G$, which ensures that no environmental degradation will occur, i.e. $S=0$. Thus one possible choice open to society is to plan for a growth path that in the long run produces zero environmental degradation.

Conditions (8) and (9) also indicate, however, that as long as some (net) environmental degradation is continuing to occur, environmental quality will decline. Equation (9) suggests that there is a lower limit to environmental quality. As noted above, if X is driven below $\underline{X}$, environmental degradation will have destroyed the natural clean-up and regenerative processes in the environment. This is tantamount to an environmental 'collapse', and economic growth leading to such a collapse can be said to be environmentally 'unsustainable'. Nevertheless, there may be conditions under which society may have no choice but opt for an unsustainable growth path.

However, in general, there will also be conditions leading society to a sustainable growth path. The pursuit of sustainability inevitably involves some intertemporal trade-offs between levels of consumption and environmental quality. For example, in Figure 2.4, the C–X curve traces for every level of environmental quality the 'sustainable' level of consumption that just leaves environmental degradation unchanged. Thus if at a given level of environmental quality, X_0, one consumes less than the sustainable level of consumption, C_0, then the environment will improve and society will be able to sustain a higher level of consumption in the future. Hence, there is, in this intertemporal sense, a positive relationship between increases in consumption (i.e. growth) and improvements in the environment.

However, the convexity of the C–X locus indicates that, as society sacrifices consumption now, so the improvements in the environment get smaller and smaller. On the other hand, the value of that sacrifice increases because of a diminishing marginal rate of substitution between consumption and environmental goods (i.e. the convexity of the indifference curve). Presumably then there would be a point where the two would balance out. In the long run equilibrium, the utility value of a unit of consumption sacrificed today should equal the discounted present value of the higher consumption and environmental quality afforded in perpetuity to future generations. The higher the discount rate, the less the latter would be and so the equilibrium would be at a lower point on the C–X curve. This is because the benefits of a unit of consumption in terms of the higher X value it permits will decline as C falls.

We therefore examine further the optimal conditions leading to sustainable versus unsustainable economic growth. Given a positive rate of time preference, r, the planning problem is to find solutions which will

$$\max \int_0^\infty e^{-rt}\, U(C, X)\, dt \tag{11}$$

$$\text{subject to } \dot{X} = -ah\,(C, X),$$

$$X(t{=}0) = X_0, \; X(t = \infty)\text{ free}, \; X \geq \underline{X}.$$

Given the continuous function $P(t)$, the Hamiltonian of the problem is:

$$H = e^{-rt}\, \{U(C, X) + P[-ah(C, X)]\}. \tag{12}$$

The first-order conditions for an interior solution are:

$$dH/dC = U_c - Pah_c = 0, \tag{13}$$

$$\text{or } P = U_c/ah_c > 0,$$

and

$$\dot{P} - rP = -dH/dX = -U_x + Pah_x, \tag{14}$$

$$\text{or } \dot{P} = [r + ah_x] \, P - U_x,$$

and

$$\dot{X} = - ah \, (C, X) . \tag{15}$$

$P(t)$ is the costate variable, which can be interpreted as the social value, or shadow price, of environmental quality. Condition (13) gives C as an explicit function of P and X with:

$$\frac{dC}{dP} = \frac{ah_c}{U_{cc} - Pah_{cc}} < 0, \tag{16}$$

and

$$\frac{dC}{dX} = \frac{Pah_{cx}}{U_{cc} - Pah_{cc}} > 0, \tag{17}$$

if $h_{cx} < 0$. As will be discussed below, the latter is important with respect to the comparative static analysis of the equilibrium.

From (14) and (15), the behaviour of the system from any initial point, (X_0, P_0), is governed by:

$$\dot{P} \gtreqless 0 \text{ as } [r + ah_x]P \gtreqless U_x, \tag{18}$$

$$\dot{X} \gtreqless 0 \text{ as } h(C, X) \gtreqless 0. \tag{19}$$

The slopes of the stationary loci satisfying these equations are given by:

$$\frac{dP}{dX} \Big|_{\dot{P}=0} = \frac{U_{xx} - Pah_{xx} - Pah_{xc} \, dC/dX}{[r + ah_x] + ah_{xc} \, dC/dP} \tag{20}$$

and

$$\frac{dP}{dX} \Big|_{\dot{X}=0} = \frac{- [h_x + h_c dC/dX]}{h_c \, dC/dP}, \tag{21}$$

which cannot be definitely signed. One possible configuration of the phase diagrams satisfying (18)–(20) is depicted in Figure 2.5, assuming $h_x \rightarrow -\infty$ as $X \rightarrow \underline{X}^+$, as seems very plausible. In Figure 2.5, (X_2^*, P_2^*) is a stable equilibrium, whereas (X_1^*, P_1^*) is unstable. If $X_0 > X_1^*$, then the optimal policy is to select P_0 so as to place the economy on a growth path that ends at the stable equilibrium (X_2^*, P_2^*). This represents an environmentally 'sustainable' growth path, given the assumption that if $\dot{X} = 0$ and $X > \underline{X}$, then biophysical

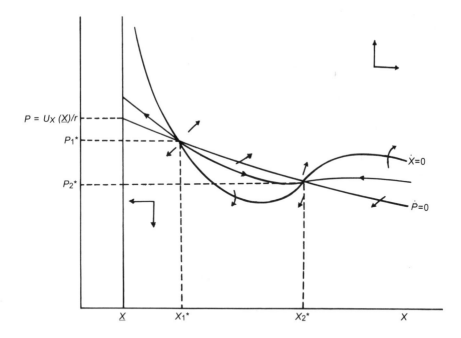

Figure 2.5 Dual equilibria solution to phase diagram.

constraints are being observed. If $X_0 = X_1^*$, then it is optimal to remain at X_1^* forever. If $X_0 < X_1^*$, then the growth path of the economy could lead to $\underline{X}$. However, this growth path is unsustainable, for at $\underline{X}$, the assimilative and regenerative capacity of the environment will have been destroyed, and the economy will be forced to consume existing internal resource stocks. Eventually, the latter will be consumed and the economy will collapse. Thus X_1^* can be considered the minimum initial level of environmental quality required to ensure a sustainable growth path.

Thus with a low initial level of environmental quality, environmentally unsustainable economic growth may be an optimal strategy. Since the benefits of increased consumption occur in the present whereas environmental degradation and collapse is a future problem, this strategy is made optimal by a high rate of discount on future utility. Consequently, both the initial level of environmental quality as well as the rate of social discount are significant factors in determining the optimal choice between sustainable and unsustainable growth as one would expect.

These intuitive results have been shown to be correct if the relationship between the rate of environmental degradation, consumption and the stock of environmental assets takes a particular form. We have assumed, as seems normal, that other things being equal the rate of degradation increases as the level of consumption increases and as the stock of environmental assets declines. In addition, however, we are required to assume that the increase in the rate of degradation as consumption increases is higher with smaller environmental stocks than with larger ones. This is the economic–ecological interpretation of the requirement that $h_{cx} < 0$. Note that the latter is a sufficient condition for the stable equilibrium to occur with a larger environmental stock at a lower discount rate, and for the unstable equilibrium to occur at a smaller environmental stock in the same circumstances.

For example, it is apparent from (18) that an increase in the discount rate would have the effect of shifting down the $\dot{P} = 0$ curve. As shown in Figure 2.6, the end result may be a unique equilibrium, but one that is stable only if $X_0 \geq X_3^*$. However, if $X_0 < X_3^*$, it may be optimal to choose an unsustainable growth path; i.e. one that heads towards $\underline{X}$. Note that, as $X_3^* > X_1^*$, an economy with an increased discount rate requires a higher minimum initial level of environmental quality to avoid a growth path that might be environmentally unsustainable. In contrast, lower discount rates would shift the $\dot{P} = 0$ curve up, requiring a lower minimum initial

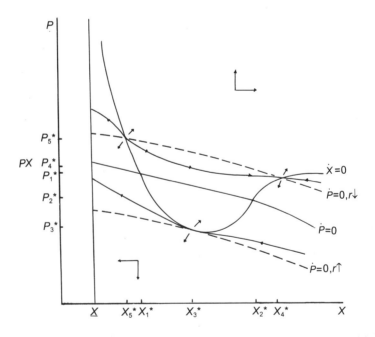

Figure 2.6 The effects of changes in *r*.

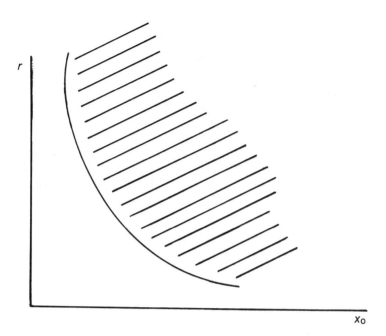

Figure 2.7 The influence of X_0 on r.

level of environmental quality to ensure sustainable growth (i.e. $X_5^* < X_1^*$). These results appear to confirm the conclusions discussed above of the role of discount rates in determining the sustainability of the economic process.

It can be confirmed that the minimum bound on the social rate of time preference, r, is not independent of the historically given level of environmental quality, X_0. Note that in (18), for $P = 0$ it is a requirement that $r > -ah_x$. Given the properties of $h(C,X)$ outlined in (7), a lower initial X will cause $-ah_x$ to rise, thus requiring a higher r to keep $P = 0$. Conversely, a higher X_0 will have a lower rate of discount. As shown in Figure 2.7, therefore, the initial level of environmental quality imposes a lower limit on the choice of r.

2.4 CONCLUSIONS

In this chapter, we have developed a model of an economic – environmental system under conditions where natural resources and ecological processes are essential to human welfare. Thus the 'sustainability' of the system is dependent on whether certain biophysical constraints are observed by the economic process. The failure to observe these constraints will lead to 'unsustainable' growth, and the system will collapse in the long run. The decision-makers affecting this choice are the current resource users in the economic–environment system today.

The results of the above model indicate that both the initial level of environmental quality as well as the rate of time preference are significant factors in determining the optimal choice between sustainable and unsustainable growth. For example, if technical condition $h_{cx} < 0$ is satisfied, with a low initial level of environmental quality and a high rate of social discount, environmentally unsustainable economic growth may be an optimal strategy as the benefits of increased consumption occur in the present whereas environmental degradation and collapse is a future problem. Moreover, the initial level of environmental quality influences choices of discount rates, so that a historically lower initial level of environmental quality leads to a high rate of discount and vice versa.

In other words, a low initial level of environmental quality forces resource users to discount the future heavily. That is, poor people faced with marginal environmental conditions have no choice but to opt for immediate economic benefits at the expense of the long-run sustainability of their livelihoods. This particularly holds for the marginal lands of the Third World, which are areas characterized not only by lower quality and productivity but also by their greater instability, especially as regards to micro-climatic, agro-ecological and soil conditions.[11]

Thus if economic development is to offer the resource-poor the opportunity of sustainable and secure livelihoods, then sustainable resource management must become a primary development goal. Many of the subsequent chapters in this volume reinforce the latter theme, with examples of dryland management and pastoral systems, biodiversity conservation and forest systems and the ecology of island economies.

REFERENCES

Barbier, E.B. (1987) The concept of sustainable economic development. *Environmental Conservation* **14**, 101–10.

Barbier, E.B. (1989a) *Economics, Natural-Resource Scarcity and Development: Conventional and Alternative Views*, Earthscan, London.

Barbier, E.B. (1989b) Sustainable agriculture on marginal land: A policy framework. *Environment*, **31**(9), 2–17 and 36–40.

Barbier, E.B. (1990a) Alternative approaches to economic–environmental interactions. *Ecological Economics*, **2**, 7–26.

Barbier, E.B., (1990b) The farm-level economics of soil erosion: The uplands of Java. *Land Economics*, **66**(9), 199–211.

Barbier, E.B. and Markandya, A. (1990) The conditions for achieving environmentally sustainable development. *European Economic Review*, **34**, 659–69.

Becker, R.A. (1982) Intergenerational equity: the capital–environment trade-off, *Journal of Environmental Economics and Management*, **9**, 165–85.

---

[11] See, for example, Barbier (1989b); Conway and Barbier (1990); FAO (1990); Leonard *et al.* (1989) and Pretty *et al.* (1992) for further discussion of this issue.

Clark, C. (1976) *Mathematical Bioeconomics: The Optimal Management of Renewable Resources*, John Wiley, New York.

Common, M. and Perrings, C. (1992) Towards an ecological economics of sustainability. Draft paper, Centre for Resource and Environmental Studies, Australia National University, Canberra.

Conway, G.R. and Barbier, E.B. (1990) *After the Green Revolution: Sustainable Agriculture for Development*, Earthscan, London.

Daly, H.E. (1987) The economic growth debate: what some economists have learned but many have not. *Journal of Environmental Economics and Management* **14**, 323–36.

d'Arge, R.C. and Kogiku, K.C. (1973) Economic growth and the environment. *Review of Economic Studies*, **40**, 61–78.

Dasgupta, P.S. (1982) *The Control of Resources*, Basil Blackwell, Oxford.

Dasgupta, P.S. and Heal, G.M. (1979) *Economic Theory and Exhaustible Resources*, Cambridge University Press, Cambridge.

FAO (1990) Sustainable development and natural resource management, Part Three in *The State of Food and Agriculture, 1989*, FAO, Rome.

Forster, B.A. (1973) Optimal consumption planning in a polluted environment. *Economic Record*, **49**, 534–45.

Forster, B.A. (1975) Optimal pollution control with a nonconstant exponential rate of decay. *Journal of Environmental Economics and Management*, **2**, 1–6.

Hartwick, J.M. (1977) Intergenerational equity and the investing of rents from exhaustible resources. *American Economic Review*, **66**, 972–4.

IUCN/UNEP/WWF (1991) *Caring for the Earth: A Strategy for Sustainable Living*, Earthscan, London.

Klaassen, G.A.J. and Opschoor, J.B. (1991) Economics of sustainability or the sustainability of economics: different paradigms. *Ecological Economics* **4**(2), 93–115.

Kneese, A.V., Ayres, R.U. and d'Arge, R.C. (1970) *Economics and the Environment: A Material Balances Approach*, Johns Hopkins University Press, Baltimore.

Leonard, H.J., with Yudelman, M., Stryker, J.D., Browder, J.O., De Boer, A.J., Campbell, T. and Jolly, A. (1989) *Environment and the Poor: Development Strategies for a Common Agenda*, Transaction Books, New Brunswick.

Mäler, K-G. (1974) *Environmental Economics: A Theoretical Inquiry*, Johns Hopkins University Press, Baltimore.

Page, T. (1977) *Conservation and Economic Efficiency: An Approach to Materials Policy*, Johns Hopkins University Press, Baltimore.

Pearce, D.W. (1987) Foundations of an ecological economics. *Ecological Modelling* **38**, 9–18.

Pearce, D.W. (1988) 'Optimal prices for sustainable development', in *Economic Growth and Sustainable Environments*, (eds D. Collard, D. Pearce and D. Ulph), Macmillan, London.

Pearce, D.W., Barbier, E.B. and Markandya, A. (1990) *Sustainable Development: Economics and Environment in the Third World*, Edward Elgar, London.

Pearce, D.W., Markandya, A. and Barbier, E.B. (1989) *Blueprint for a Green Economy*, Earthscan, London.

Pezzey, J. (1989) *Economic Analysis of Sustainable Growth and Sustainable Development*. Environment Department Working Paper No. 15, The World Bank, Washington DC.

Pretty, J. Guijt, I., Scoones, I. and Thompson, J. (1992) Regenerating agriculture: The agroecology of low-external input and community-based development, in *Policies for a Small Planet* (ed. J. Holmberg), Earthscan, London.

Smith, V.L. (1977) Control theory applied to natural and environmental resources: An exposition. *Journal of Environmental Economics and Management*, **4**, 1–24.

Solow, R.M. (1986) On the intertemporal allocation of natural resources. *Scandinavian Journal of Economics*, **88**(1), 141–9.
UNCED Secretariat (1992) *Agenda 21*, UNCED, New York.
Victor, P.A. (1972) *Pollution, Economics and the Environment*, Allen and Unwin, London.
World Commission on Environment and Development (1987) *Our Common Future*, Oxford University Press, Oxford.

3

Ecological economic systems analysis: order and chaos

Robert Costanza

3.1 WHAT IS SYSTEMS ANALYSIS?

Systems analysis is the study of systems, groups of interacting, interdependent parts linked together by complex exchanges of energy, matter, and information. There is a key distinction between 'classical' science and system science. Classical (or reductionist) science is based on the resolution of phenomena into isolatable causal trains and the search for basic, 'atomic' units or parts of the system. Classical science depends on weak or non-existent interaction between parts and essentially linear relations among the parts, so that the parts can be added together to give the behaviour of the whole. These conditions are not met in the entities called systems. A 'system' is characterized by strong (usually non-linear) interactions between the parts, feedbacks (making resolution into isolatable causal trains difficult or impossible) and the inability to simply 'add-up' small-scale behaviour to arrive at large-scale results (von Bertalanffy, 1968). Ecological and economic systems obviously exhibit these characteristics of systems, and are not well understood using the methods of classical, reductionist science.

While almost any subdivision of the universe can be thought of as a 'system,' systems analysts usually look for boundaries that minimize the interaction between the system under study and the rest of the universe in order to make their job easier. Some claim that nature presents a convenient hierarchy of scales based on these labour-saving boundaries, ranging from

Economics and Ecology: New frontiers and sustainable development.
Edited by Edward B. Barbier. Published in 1993 by Chapman & Hall, 2–6 Boundary Row, London SE1 8HN. ISBN 0 412 48180 4.

atoms to molecules to cells to organs to organisms to populations to communities to ecosystems (including economic, and/or human domi- nated ecosystems) to bioregions to the global system and beyond (Allan and Starr, 1982; O'Neill *et al.*, 1986). By studying the similarities and differences between different kinds of systems at different scales and resolutions, one can develop hypotheses and test them against other sys- tems to explore their degree of generality and predictability.

One might define systems analysis as the scientific method applied both across and within disciplines, scales, resolutions, and system types. In other words, it is an integrative manifestation of the scientific method, while most of the traditional or classical scientific disciplines tend to dissect their subjects into smaller and smaller parts hoping to reduce the problem to its essential elements. Beyond this distinction between synthesis and reduc- tion, systems analysis usually has connotations of mathematical modelling applied to these integrative problems. While this is neither a necessary nor sufficient condition for systems analysis, it is a common characteristic, if for no other reason than that systems tend to be complex and mathematical modelling (especially on computers) is usually necessary to handle that complexity.

3.1.1 A short history of systems analysis

Because of this practical requirement for computer simulation in order to handle even fairly simple systems, the history of systems analysis is tightly linked with the history of the computer. While computers first appeared in the 1950s, their widespread use did not commence until the 60s and 70s, and did not become universal and common until the 1980s. With the increasing availability, power, and 'user-friendliness' of computers has come an increasing feasibility of systems analysis. Today, anyone with a few thousand dollars can buy a personal computer along with some relevant software and begin to do practical systems analysis.

The possibility for this sort of analysis was recognized and early practical models were developed more or less independently by workers in econom- ics, ecology, industrial management, and what was then called cybernetics. Early 'systems analysts' in economics include Leontief (1941) and Von Neumann and Morgenstern (1953), mainly focused on static input–output networks and games. In ecology, Odum (1971), Patten (1971–1976) and Hannon (1973) were among the early practitioners of both dynamic com- puter simulation and static network analysis. The International Biosphere Program (IBP) was an early large-scale attempt to perform ecological systems analysis for a range of ecosystems (Innis, 1978).

Jay Forester of MIT began modelling complex industrial systems in the early 60s (Forrester, 1961) and has spawned one of the most prolific schools of systems analysis. One milestone in this school's development was the world systems model reported in *Limits to Growth* (Meadows *et al.*, 1972) and subsequent criticisms (Cole *et al.*, 1973; Oltmans, 1974) and expansions (Mesarovic and Pestel, 1974; Ehrlich and Holdren, 1988; Pestel, 1989; Meadows *et al.*, 1992). But Ludwig von Bertalanffy is usually acknowledged as the father of systems analysis (von Bertalanffy, 1950). His seminal book titled *General Systems Theory* (von Bertalanffy, 1968) laid out the general principles of systems analysis sketched above, that still hold up well today.

As mentioned earlier, operationalizing these principles and actually doing practical systems analysis has been linked to the development of the computer and to the development of large computerized data bases to feed the models. According to von Bertalanffy (1968:p18) 'the system problem is essentially the problem of the limitations of analytical procedures in science'. Recent years have seen an explosion in our ability to overcome these limitations and to actually study the complex, non-linear, scale-dependent behaviour of systems.

One convenient way (but certainly not the only way) to split the work in systems analysis into categories is to differentiate between analyses that attempt to examine the structure of complex systems (I'll call it **network analyses**) and those that examine the dynamics of complex systems (I'll call it **simulation modelling**). While these applications are related, their goals and assumptions are fairly distinct and I think it is helpful to talk about them separately.

3.1.2 Network analysis: uncovering structure in complex systems

Ecology is often defined as the study of the relationships between organisms and their environment. The quantitative analysis of interconnections between species and their abiotic environment has therefore been a central issue. The mathematical analysis of interconnections is also important in several other fields. Practical quantitative analysis of interconnections in complex systems began with the economist Wassily Leontief (1941) using what has come to be called Input–Output (I–O) Analysis. More recently, these concepts (also sometimes called Flow Analysis) have been applied to the study of interconnections in ecosystems (Hannon, 1973, 1976, 1979, 1985a,b,c; Costanza and Neill, 1984). Related ideas were developed from a different perspective in ecology, under the heading of Compartmental Analysis (Barber, Patten and Finn, 1979; Finn, 1976; Funderlic and Heath, 1971; Hett and O'Neill, 1971). Isard was the first to attempt combined ecological economic system I–O analysis (Isard, 1972) and combined ecological economic mass balance models have been proposed by several other authors (Daly, 1968; Ayres and Kneese, 1969; Cumberland, 1987). We refer to the total of all variations of the analysis of ecological and/or economic networks as network analysis.

Network analysis holds the promise of allowing an integrated, quantitative, hierarchical treatment of all complex systems, including ecosystems and combined ecological economic systems. One promising route is the use of 'ascendancy' (Ulanowicz 1980, 1986) and related measures (Wulff, Field and Mann, 1989) to measure the degree of organization in ecological, economic, or any other networks. Measures like ascendancy go several steps beyond the traditional diversity indices used in ecology. They estimate not only how many different species there are in a system but, more importantly, how those species are organized. This kind of measure may provide the basis for a quantitative and general index of system health applicable to both ecological and economic systems.

Another promising avenue of research in network analysis has to do with its use for 'pricing' commodities in ecological or economic systems. The 'mixed units' problem arises in any field that tries to analyse interdependence in complex systems that have many different types and qualities of interacting commodities. Ecology and economics are two such fields. Network analysis in ecology has avoided this problem in the past by arbitrarily choosing one commodity flowing through the system as an index of interdependence (i.e. carbon, enthalpy, nitrogen, etc.). This ignores the interdependencies between commodities and assumes that the chosen commodity is a valid 'tracer' for relative value or importance in the system. This assumption is unrealistic and severely limits the comprehensiveness of an analysis whose major objective is to deal comprehensively with whole systems.

There are evolving methods for dealing with the mixed units problem based on analogies to the calculation of prices in economic input–output models. Starting with a more realistic **commodity by process** description of ecosystem networks that allows for joint products one can use **energy intensities** to ultimately convert the multiple commodity description into a pair of matrices that can serve as the input for standard (single commodity) network analysis. The new single commodity description incorporates commodity and process interdependencies in a manner analogous to the way economic value incorporates production interdependencies in economic systems (Costanza and Hannon, 1989). This analysis would allow 'systems' valuation of components of ecosystems and combined ecological and economic systems as a complement to subjective evaluations.

3.1.3 Computer simulation modelling: the dynamics of complex systems

Modelling the dynamics of complex systems has a long and interesting history. While much of this work goes back well before computers, I shall concentrate only on computer simulation modelling. Until computers came into play, the equations that described the dynamics of systems had to be solved analytically. This severely limited the level of complexity of the

Table 3.1 The limits of analytical methods in solving mathematical problems (from von Bertalanffy, 1968). The thick solid line divides the range of problems that are solvable with analytical methods from those that are very difficult or impossible using analytical methods and require numerical methods and computers to solve. Almost all 'systems' problems fall in the range that requires numerical methods

Equations	Linear			Non-linear		
	One equation	Several equations	Many equations	One equation	Several equations	Many equations
Algebraic	trivial	easy	essentially impossible	very difficult	very difficult	impossible
Ordinary differential	easy	difficult	essentially impossible	very difficult	impossible	impossible
Partial differential	difficult	essentially impossible	impossible	impossible	impossible	impossible

systems that could be studied, and the complexity of the dynamics that could be studied for any particular system. Table 3.1 shows the limits of analytical methods in solving various classes of problems.

As Table 3.1 shows, only relatively simple linear systems of algebraic or differential equations can be analysed analytically. The problem is that most complex, living systems (like economies and ecosystems) are decidedly non-linear and efforts to approximate their dynamics with linear equations have been of only very limited usefulness.[1] In addition, complex systems exhibit discontinuous and sometimes chaotic behavior (Rosser, 1991) that can only be adequately represented with computer simulations.

The field of computer simulation has been growing so rapidly in recent years that it would be impossible to provide even a brief overview of all the kinds of models that have been created and effectively used. I'll concentrate instead on ecosystem models as an example, and on some general hypotheses and principles of modelling complex systems that have developed from these models. Ecosystem models can be differentiated from population models in that the former include whole ecosystems (both the biotic and abiotic components) while the latter include only populations of organisms. Ecosystem models tend to be more complex and realistic while population models tend to be more general and simple. The majority of ecosystem models in the literature are designed to predict dynamic behaviour while treating the system as spatially homogeneous (Costanza and Sklar, 1985). Many existing ecosystem models are process-based, in the sense of attempting to mimic (at least in a very aggregated way) the underlying physical and

---

[1] I differentiate here between the use of linear systems of equations to model complex system dynamics (which I think does not work well) vs. the use of linear systems to understand structure (which I think does work reasonably well) as in the network analysis described above. Integrating these views of structure and dynamics is a key item on the research agenda.

ecological processes in the system, as opposed to statistical or probabilistic models which are based directly on observed correlations in the data, generally without specifying mechanisms. One way to extend this process-based approach to model spatial dynamics is to arrange a spatial array of point ecosystem models and connect them with fluxes of, for example, water, nutrients, etc. and with appropriate rules to govern successional, evolutionary, or other changes in the structure of the system. This approach is somewhat analogous to that employed in General Atmospheric Circulation Models (GACMs) used in long-term climate modelling (Williams, Barry and Washington, 1974; Washington and Williamson, 1977; Potter *et al.*, 1979; Schlesinger and Zhao, 1989), but it also incorporates elements of cellular automata and expert systems modelling incorporated in the successional rules. For example, we developed the Coastal Ecological Landscape Spatial Simulation (CELSS) model, consisting of 2479 1 km^2 spatial cells to simulate a rapidly changing section of the Louisiana coast and predict long-term (50 to 100 year) spatially articulated changes in this landscape as a function of various management alternatives and natural and human-influenced climate variations (Costanza, Sklar and White, 1990).

The model was run on a CRAY supercomputer from initial conditions in 1956 through 1978 and 1983 (years for which additional data was available for calibration and validation) and on to the year 2033 with a maximum of weekly time steps. It accounted for 89.6% of the spatial variation in the 1978 calibration data and 79% of the variation in the 1983 verification data. Various future and past scenarios were analysed with the model, including the future impacts of various Atchafalaya River levee extension proposals, freshwater diversion plans, marsh damage mitigation plans, future global sea level rise and the historical impacts of past human activities and past climate patterns.

3.1.4 Resolution and predictability

Experimenting with the CELSS model raised some interesting questions about the influence of resolution (including spatial, temporal, and complexity) on the performance of models, in particular their predictability. We analysed the relationship between resolution and predictability and found that while increasing resolution provides more descriptive information about the patterns in data it also increases the difficulty of accurately modelling those patterns. There are limits to the predictability of natural phenomena at particular resolutions, and 'fractal like' rules that determine how both 'data' and 'model' predictability change with resolution.

To test these ideas we analysed land use data by resampling map data sets at several different spatial resolutions and measuring predictability at each. Predictability (Colwell, 1974) measures the reduction in uncertainty (scaled on a 0–1 range) about one variable given knowledge of others using categorical data. We defined spatial **auto-predictability** (P_a) as the reduction in uncertainty about the state of a pixel in a scene, given knowledge of the state of adjacent pixels in that scene, and spatial **cross-predictability** (P_c) as the reduction in uncertainty about the state of a pixel in a scene given knowledge of the state of corresponding pixels in other scenes. P_a is a measure of the internal pattern in the data, while P_c is a measure of the ability of some other model to represent that pattern.

We found a strong linear relationship between the log of P_a and the log of resolution (measured as the number of pixels per square kilometre). This fractal-like characteristic of 'self-similarity' with decreasing resolution implies that predictability, like the length of a coastline, may be best described using a unitless dimension that summarizes how it changes with resolution. We defined a 'fractal predictability dimension' (D_P) in a manner analogous to the normal fractal dimension (Mandelbrot, 1977, 1983) that summarizes this relationship. The D_P allows convenient scaling of predictability measurements taken at one resolution to any other.

Cross-predictability (P_c) can be used for pattern matching and testing

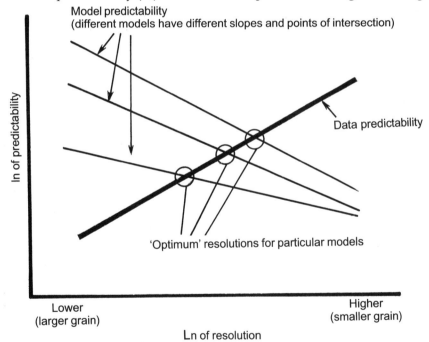

Figure 3.1 Hypothetical relationship between resolution and predictability of data and models. Data predictability is the degree to which the uncertainty about the state of landscape pixels is reduced by knowledge of the state of adjacent pixels in the same map. Model predictability is the degree to which the uncertainty about the state of pixels is reduced by knowledge of the corresponding state of pixels in output maps from various models of the system.

the fit between scenes. In this sense it relates to the predictability of models versus the inherent predictability in the data revealed by P_a. While P_a generally increases with increasing resolution (because more information is being included), P_c generally falls or remains stable (because it is easier to model aggregate results than fine grain ones). Thus we can define an optimal resolution for a particular modelling problem that balances the benefit in terms of increasing data predictability (P_a) as one increases resolution, with the cost of decreasing model predictability (P_c). Figure 3.1 shows this relationship in generalized form.

These results may be generalizable to all forms of resolution (spatial, temporal, and number of components) and may shed some interesting light on the debate over 'chaotic' systems. Chaos, in this view, is low model predictability (for longer term predictions) that occurs as a natural consequence of high resolution. Lowering the resolution can increase the model predictability by averaging out much of the chaotic behaviour, at the expense of losing detail about the phenomenon.

3.1.5 Toward integrated, multi-scale, transdisciplinary, and pluralistic modelling

Richard Levins (1966) first described the fundamental trade-offs in modelling between realism, precision, and generality. No single model can maximize all three of these goals and the choice of which objectives to pursue depends on the fundamental purposes of the modelling study.

Adequately evaluating the dynamics of complex systems requires a pluralistic approach (Norgaard, 1989; Rapport, 1989) and an ability to integrate and synthesize the many different perspectives that can be taken. There is probably not one right approach or paradigm, because, like the blind men and the elephant, the subject is too big and complex to touch it all with one limited set of perceptual tools. Rather, we need to extend our view to cover the pluralism of approaches that may shed light on the problem, and also develop the ability to use all of the available light to view and understand the system.

We need an integrated, multi-scale, transdisciplinary, and pluralistic (IMTP), approach to quantitative modelling of systems (including organisms, ecosystems, and ecological economic systems). While this approach has frequently been suggested (cf. Norgaard, 1989), it is difficult to operationalize with traditional funding mechanisms.

The IMTP approach would allow the relationships between scales and modelling approaches to be directly investigated, and would result in a deeper understanding of the systems under study. It would produce new ways of **scaling**, or using information at one scale to build models at other scales.

Predicting these kinds of ecosystem impacts requires sophisticated computer simulation models that represent a synthesis of the best available understanding of the way these complex systems function (Costanza *et al.*, 1990). Development of this modelling capability is essential for regional ecosystem management and also for modelling regional and global ecosystem response to regional and global climate change, sea level rise resulting from atmospheric CO_2 enrichment, acid precipitation, toxic waste dumping, and a host of other potential impacts.

Several recent developments make this kind of modelling feasible, including the improved accessibility of extensive spatial and temporal data bases from remote sensing, historical aerial photography, and other sources (including EPA's new EMAP programme), and advances in computer power and convenience that make it possible to build and run predictive models at the necessary levels of spatial and temporal resolution (Costanza *et al.*, 1990).

3.2 SCIENTIFIC UNCERTAINTY: BEYOND RISK, INTO THE ABYSS

One of the primary reasons for the problem is the issue of scientific uncertainty. Not just its existence, but the radically different expectations and modes of operation that science and policy have developed to deal with it. If we are to solve this problem, we must understand and expose these differences about the nature of uncertainty and design better methods to incorporate it into the policy making process.

To understand the scope of the problem, it is necessary to differentiate between risk (which is an event with a known probability) and true uncertainty (which is an event with an unknown probability). Every time you drive your car you run the risk of having an accident, because the probability of car accidents is known with very high certainty. We know the risk involved in driving because, unfortunately, there have been many car accidents on which to base the probabilities. These probabilities are known with enough certainty that they are used by insurance companies to set rates that will assure those companies of a certain profit. There is little uncertainty about the risk of car accidents.

If you live near the disposal site of some newly synthesized toxic chemical you may be in danger as well, but no one knows to what extent. No one knows even the probability of your getting cancer or some other disease from this exposure, so there is true uncertainty. Most important environmental problems suffer from true uncertainty, not merely risk.

Science treats uncertainty as a given, a characteristic of all information that must be honestly acknowledged and communicated. Over the years scientists have developed increasingly sophisticated methods to measure and communicate uncertainty arising from various causes. It is important to note that the progress of science has, in general, uncovered more uncertainty rather than leading to the absolute precision that the lay public often mistakenly associates with 'scientific' results. The scientific method can only set boundaries on the limits of our knowledge. It can define the edges of the envelope of what is possible, but often this envelope is very large and the shape of its interior can be a complete mystery. Science can tell us what the range of uncertainty is and about global warming and toxic chemicals, and maybe something about the probabilities of different outcomes, but in many important cases it cannot tell us which of the possible outcomes will occur with any accuracy.

Policy making, on the other hand, abhors uncertainty and gravitates to the edges of the envelope. The reasons for this are clear. The goal of policy is making clear, defensible decisions, often codified in the form of laws and regulations. While legislative language is often open to interpretation, regulations are much easier to write and enforce if they are stated in clear, black and white, absolutely certain terms. For most of criminal law this works well. Either Mr. Cain killed his brother or he didn't and the only question is whether there is enough evidence to prove it beyond a shadow of a doubt (i.e. with zero uncertainty). What good would it do to conclude that there was a 80% chance that Mr. Cain killed his brother? But most scientific studies come to just these kinds of conclusions, because that is the nature of science. Science defines the envelope while the policy process gravitates to its edges – generally the edge which best advances the policy maker's political agenda. We need to deal with the whole envelope and all its implications if we are to rationally use science to make policy.

The problem is most severe in the environmental area. Building on the legal traditions of criminal law, policy makers and environmental regulators want absolute, certain, information when designing environmental regulations. Much of environmental policy is based on scientific studies of the likely health, safety and ecological consequences of human actions. Information gained from these studies is therefore only certain within their epistemological and methodological limits (Thompson, 1986). Particularly with the recent shift in environmental concerns from visible, known pollution to more subtle threats, like radon, regulators are confronted with decision making outside the limits of scientific certainty with increasing frequency (Weinberg, 1985). Problems arise when regulators ask scientists for answers to unanswerable questions. For example, the law may mandate that the regulatory agency come up with safety standards for all known toxins when little or no information is available on the impacts of many of these chemicals. When trying to enforce these regulations after they are drafted, the problem remains that it is not known (beyond a shadow of a doubt anyway) what the impacts are. It is not possible to determine if the local chemical company contributed to the death of some of the people in the vicinity of their toxic waste dump. One cannot prove the smoking–lung cancer connection in any direct, causal way (i.e. in the courtroom sense), only as a statistical relationship. Global warming may or may not happen after all.

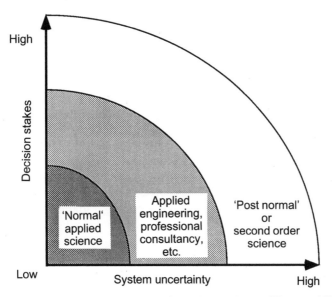

High

Decision stakes

'Normal' applied science

Applied engineering, professional consultancy, etc.

'Post normal' or second order science

Low   System uncertainty   High

Figure 3.2 Three kinds of science (from Funtowicz and Ravetz, 1991).

As they are currently set up, environmental regulations demand certainty and when scientists are pressured to supply this nonexistent commodity there is not only frustration and poor communication but mixed messages in the media as well. Because of uncertainty, environmental issues can often be manipulated for political and economic gain. The size of the stakes can often determine how uncertainty is dealt with in the political arena. The situation can be summarized as shown in Figure 3.2, with uncertainty plotted against decision stakes. It is only the area near the origin with low uncertainty and low stakes that is the domain of 'normal applied science'. Higher uncertainty or higher stakes result in a much more politicized environment. Moderate values of either correspond to 'applied engineering' or 'professional consultancy' which allows a good measure of judgement and opinion. On the other hand, current methods are not in place to deal with high values of either stakes or uncertainty, which require a new approach, what might be called 'post normal' or 'second order science' (Funtowicz and Ravetz, 1991). This 'new' science is really just the application of the essence of the scientific method to areas that have not yet had this treatment. The scientific method does not, in its basic form, imply anything about the precision of the results achieved. It does imply a forum of open and free inquiry without preconceived answers or agendas aimed at determining the envelope of our knowledge and the magnitude of our ignorance. In response to the uncertainty inherent in this class of problem, it has been suggested that they be dealt with using the 'precautionary principle'. This approach is a process of sequential decision making in which policymakers take an initially cautious approach that may be relaxed as evidence becomes more available. It 'implies the commitment of resources now to safeguard against the potentially adverse future outcomes of some decision' (Perrings, 1991).

Implementing this view of science requires a radically different, non-regulatory approach to environmental protection that acknowledges the existence of uncertainty rather than denying it, and includes a mechanism to safeguard against its potentially catastrophic effects. The real challenge is to adjust incentives so that the appropriate parties pay for uncertainty and have appropriate incentives to reduce its detrimental effects. Without this adjustment, the full costs of environmental use will continue to be left out of the costs of production and represent a free subsidy from society to those who are profiting from environmental degradation (Peskin, 1991).

3.2.1 Dealing with uncertainty

How then should we deal with the enormous uncertainty inherent in environmental issues? For starters, we should accept uncertainty and learn to communicate it better. We should also fundamentally change the approach to environmental management. In the past two decades there has been exhaustive discussion in the literature of the efficiency that can theoretically be achieved in environmental management through the use of market mechanisms. These mechanisms are designed to alter the pricing structure of the present market system to reflect better the total social cost of environmental damage by incorporating externalities into the firm's cost structure. Suggested market alternatives include pollution taxes, tradable pollution discharge permits, financial responsibility requirements and deposit–refund systems. However, uncertainty is not thoroughly addressed by market based alternatives despite its importance and pervasiveness in environmental problems.

An innovative instrument currently being researched to manage the environment under uncertainty is a **flexible environmental assurance bonding system** (Costanza and Perrings, 1990). This variation of the deposit–refund system is designed to incorporate environmental criteria and uncertainty into the market, and to induce positive environmental technological innovation. It works in this way: in addition to charging a firm directly for known environmental damages, an assurance bond equal to the current best estimate of the largest potential future environmental damages would be levied and kept in an interest-bearing escrow account for a predetermined length of time. In keeping with the precautionary principle, this system requires the commitment of resources now to offset the potentially catastrophic future effects of current activity. Portions of the bond (plus interest) would be returned if and only if the firm could demonstrate that the suspected worst case damages had not occurred or would be less than originally assessed. If damages did occur, the portions of the bond would be used to rehabilitate or repair the environment, and possibly to compensate injured parties. By requiring the users of environmental resources to post a bond adequate to cover uncertain future environmental damages (with the possibility for refunds), the burden of proof (and the cost of the uncertainty) is shifted from the public to the resource user. At the same time, firms are not charged in any final way for uncertain future damages and can recover portions of their bond in proportion to how much better their performance is than the worst case.

Deposit–refund systems, in general, are not a new concept. They have been applied to consumer policy, conservation policy, environmental policy, and other efficiency objectives. Deposit–refund systems can be market generated or government initiated and are often performance based. For example, deposit–refund systems are currently effectively used to encourage the proper management of beverage containers and used lubricating oils (Bohm, 1981).

Environmental assurance bonding would be similar to the producer-paid performance bonds often required for federal, state or local government work. For example The Miller Act (40 U.S.C. 270), a 1935 federal statute, requires contractors performing construction contracts for the federal government to secure performance bonds. Performance bonds provide a contractual guarantee that the principal (the entity which is doing work or providing service) will perform in a designated way. Often, bonds are required for work done in the private sector as well.

Performance bonds are frequently posted in the form of corporate surety bonds which are licensed under various insurance laws and, under their charter, have legal authority to act as financial guarantee for others. The unrecoverable cost of this service is usually 1–5% of the bond amount. However, under the Miller Act (FAR 28.203-1 and 28.203-2), any contract above a designated amount ($25 000 in the case of construction) can be backed by other types of securities, such as US bonds or notes, in lieu of a bond guaranteed by a surety company. In this case, the contractor provides a duly executed power of attorney and an agreement authorizing collection on the bond or notes if they default on the contract (PRC Environmental Management, 1986). If the contractor performs all the obligations specified in the contract, the securities are returned to the contractor and the usual cost of the surety is avoided.

Environmental assurance bonds would work in a similar manner (by providing a contractual guarantee that the principal would perform in an environmentally benign manner) but would be levied for the current best estimate of the largest potential future environmental damages. In most cases, these bonds could be administered by the regulatory authority that currently manages the operation or procedure. For example, in the US the Environmental Protection Agency could be the primary authority. In some cases or countries, however, it may be better to set up a completely independent agency to administer the bonds.

Protocol for worst case analysis already exists within the EPA. In 1977 the US Council on Environmental Quality required worst case analysis for implementing NEPA (National Environmental Policy Act of 1969). This required the regulatory agency to consider the worst environmental consequences of an action when scientific uncertainty was involved (Fogleman, 1987).

Strong economic incentives are provided by the bond to reduce pollution, to research the true costs of environmentally damaging activities, and to develop new innovative, cost-effective pollution control technologies. The bonding system is an extension of the 'polluter pays principle' to 'the polluter pays for uncertainty as well' or the 'precautionary polluter pays principle' (4P). It would allow a much more proactive (rather than reactive) approach to environmental problems because the bond is paid up front, before the possibility of damage is done. It would tend to foster prevention rather than cleanup by unleashing the creative resources of firms on finding more environmentally benign technologies, since these technologies would also be economically attractive. Competition in the marketplace would lead to environmental improvement rather than degradation. The bonding system would deal more appropriately with scientific uncertainty.

3.3 CONCLUSIONS

Ecological and economic systems are truly 'systems', in the sense that they have the characteristics of strong (usually non-linear) interactions between the parts, many feedbacks (making resolution into isolatable causal trains difficult or impossible) and the inability to simply 'add-up' small scale behaviour to arrive at large-scale results. Systems analysis has evolved with a pluralistic, transdisciplinary set of tools to deal with these characteristics of systems, ranging from network analysis to investigate their structure, to dynamic simulation modelling to investigate their dynamics, to hierarchy theory and multiscale analysis to investigate scaling, to economic instruments to deal with the irreducible uncertainty inherent in our understanding of complex systems in their management. The stage is set for future research on linked ecological economic systems aimed at a deeper understanding of their structure, dynamics, and scale-dependence, and the development and testing of more effective instruments for their management to assure sustainability.

ACKNOWLEDGEMENTS

This research was supported in part by NSF Grants BSR-8814272 titled: Responses of a major land margin ecosystem to changes in terrestrial nutrient inputs, internal nutrient cycling, production and export, and BSR-8906269 titled: Landscape modeling: the synthesis of ecological processes over large geographic regions and long time scales.

REFERENCES

Allen, T.F.H. and Starr, T.B. (1982) *Hierarchy*. University of Chicago Press, Chicago.

Ayres, R.U. and Kneese, A.V. (1969) Production, consumption and externalities. *American Economic Review*, **59**, 282–97.

Barber, M., Patten, B. and Finn, J. (1979) Review and Evaluation of I-O Flow Analysis for Ecological Applications, in: *Compartmental Analysis of Ecosystem Models*, Vol 10 of *Statistical Ecology*, (eds J. Matis, B. Patten and G. White), International Cooperative Publishing House, Bertonsville, Md.

Bohm, P. (1981) *Deposit-Refund Systems*, Resources for the Future, Inc., The John Hopkins University Press, Baltimore and London.

Cole, H.S.D., Freeman, C. Jahoda, M. and Pavitt, K.L.R. (eds.) (1973) *Models of Doom: a Critique of the Limits to Growth*, Universe Books, New York, NY.

Colwell, R.K. (1974) Predictability, constancy, and contingency of periodic phenomena. *Ecology*, **55**, 1148–53.

Costanza, R. and Hannon, B.M. (1989) Dealing with the 'mixed units' problem in ecosystem network analysis, in: *Network Analysis of Marine Ecosystems: Methods and Applications* (eds F. Wulff, J.G. Field, and K.H. Mann), Coastal and Estuarine Studies Series, Springer-Verlag, Heidleberg, pp. 90–115.

Costanza, R., and Neill, C. (1984) Energy intensities, interdependence, and value in ecological systems: A linear programming approach. *Journal of Theoretical Biology*, **106**, 41–57

Costanza, R. and Perrings, C. (1990) A flexible assurance bonding system for improved environmental management. *Ecological Economics*, **2**, 57–76.

Costanza, R., and Sklar, F.H. (1985) Articulation, accuracy, and effectiveness of mathematical models: A review of freshwater wetland applications. *Ecological Modelling*, **27**, 45–68

Costanza, R., Sklar, F.H. and White, M.L. (1990) Modeling coastal landscape dynamics. *BioScience*, **40**, 91–107

Cumberland, J.H. (1987) Need economic development be hazardous to the health of the Chesapeake Bay? *Marine Resource Economics*, **4**, 81–93.

Daly, H. (1968) On economics as a life science. *Journal of Political Economy*, **76**, 392–406.

Ehrlich, P.R. and Holdren, J.P. (1988) *The Cassandra Conference: Resources and the Human Predicament*, Texas A and M University Press, College Station TX.

Finn, J. (1976) The Cycling Index. *J. Theo. Biology*, **56**, 363–73

Fogleman, Valerie M. (1987) Worst case analysis: A continued requirement under the National Environmental Policy Act? *Columbia Journal of Environmental Law*, **13**, 53.

Forrester, J.W. (1961) *Industrial Dynamics*. MIT Press, Cambridge, MA.

Funderlic, R and Heath, M. (1971) Linear Compartmental Analysis of Ecosystems, Oak Ridge Natl Lab, ORNL-IBP-71-4.

Funtowicz, S.O. and Ravetz, J.R. (1991) A new scientific methodology for global environmental problems, in: *Ecological Economics: The Science and Management of Sustainability*, (ed. R. Costanza), Columbia University Press, New York, pp. 137–52.

Hannon, B. (1973) The structure of ecosystems. *J. Theo. Biology*, **41**, 535–46.

Hannon, B. (1976) Marginal product pricing in the ecosystem. *J. Theo. Biology*, **56,** 256–67.

Hannon, B. (1979) Total energy costs in ecosystems. *J. Theo. Biology*, **80,** 271–93.

Hannon, B. (1985a) Ecosystem flow analysis. *Canadian Journal of Fisheries and Aquatic Sciences*, 213, in *Ecological Theory for Biological Oceanography* (eds R. Ulanowicz and T. Platt), pp. 97–118.

Hannon, B. (1985b) Conditioning the ecosystem. *Mathematical Biology*, **75,** 23–42.

Hannon, B. (1985c), Linear dynamic ecosystems. *J. Theo. Biology*, **116,** 89–98.

Hett, J and O'Neill, R. (1971) Systems Analysis of the Aleut Ecosystem. US-IBP, Deciduous Forest Biome Memo Report, 71–16, September.

Innis, G. (1978) *Grassland Simulation Model*, Ecology studies No. 26, Springer-Verlag, New York.

Isard, W. (1972) *Ecologic-Economic Analysis for Regional Development*. The Free Press, New York.

Leontief, W. (1941) *The Structure of American Economy, 1919–1939,* Oxford University Press, New York.

Levins, R. (1966) The strategy of model building in population biology. *American Scientist*, **54,** 421–31.

Mandelbrot, B.B. (1977) *Fractals. Form, Chance and Dimension*. W.H. Freeman and Co., San Francisco, CA.

Mandelbrot, B.B. (1983) *The Fractal Geometry of Nature*, W.H. Freeman and Co., San Francisco, CA.

Meadows, D.H., Meadows, D.L., Randers, J. and Behrens, W.W. (1972) *The Limits to Growth*, Universe, New York.

Meadows, D.H., Meadows, D.L. and Randers, J. (1992) *Beyond the Limits: Confronting Global Collapse, Envisioning a Sustainable Future*, Chelsea Green, Post Mills, VT.

Mesarovic, M. and Pestel, E. (1974) *Mankind at the Turning Point: the Second Report to the Club of Rome*, Dutton, New York, NY

Norgaard, R.B. (1989) The case for methodological pluralism. *Ecological Economics*, **1,** 37–57.

Odum, H.T. (1971) *Environment, Power and Society*, John Wiley, New York, NY

Oltmans, W.L. (1974) *On Growth*, Capricorn, New York.

O'Neill, R.V., DeAngelis, D.L., Waide, J.B. and Allen, T.F.H. (1986) *A Hierarchial Concept of Ecosystems*, Princeton University Press, Princeton, NJ.

Patten, B.C. (1971–1976) *Systems Analyses and Simulation in Ecology*, Vols. 1–4. Academic Press, New York, NY.

Perrings, C. (1991) Reserved rationality and the precautionary principle: Technological change, time and uncertainty in environmental decision making, in: *Ecological Economics: the Science and Management of Sustainability*, (ed. R. Costanza), Columbia University Press, New York, pp. 153–66.

Peskin, Henry M. (1991) Alternative environmental and resource accounting approaches, in: *Ecological Economics: the Science and Management of Sustainability*, (ed R. Costanza), Columbia University Press, New York, pp. 176–93.

Pestel, E. (1989) *Beyond the Limits to Growth: a Report to the Club of Rome*, Universe Books, New York, NY.

Potter, G.L., Ellsaesser, H.W., MacCracken, M.C. and Luther, F.M. (1979) Performance of the Lawerce Livermore Laboratory zonal atmospheric model, in: *Report of the JOC Study Conference on Climate Models: Performance, Intercomparison and Sensitivity Studies*, (ed. W.L. Gates), Global Atmospheric Research Programme Series No. 22, Washington, DC, pp 852–71.

PRC Environmental Management. (1986) *Performance Bonding*. A final report prepared for the U.S. Environmental Protection Agency, Office of Waste Programs and Enforcement, Washington, DC.

Rapport, D.J. (1989) What constitutes ecosystem health? *Perspectives in Biology and Medicine*, **33**, 120–32.

Rosser, J.B. (1991) *From Catastrophe to Chaos: A general theory of economic discontinuities*, Kluwer, Dordrecht.

Schlesinger, M.E. and Zhao, Z.C. (1989) Seasonal climatic changes induced by doubled CO_2 as simulated by the OSU atmospheric GCM/mixed-layer ocean model. *Journal of Climate*, **2**, 463–99.

Thompson, P.B. (1986) Uncertainty arguments in environmental issues. *Environmental Ethics*, **8**, 59–76.

von Bertalanffy, L. (1950) An outline of general system theory. *Brit. J. Philos. Sci.*, **1**, 139–64.

von Bertalanffy, L. (1968) *General System Theory: Foundations, Development, Applications*, George Braziller, New York, NY.

Von Neumann, J. and Morgenstern, O. (1953) *Theory of Games and Economic Behavior*, Princeton University Press, Princeton, NJ.

Ulanowicz, R.E. (1980) An hypothesis on the development of natural communities. *J. Theor. Biol.* **85**, 223–45.

Ulanowicz, R.E. (1986) *Growth and Development: Ecosystems Phenomenology*. Springer-Verlag, NY.

Washington, W.M. and Williamson, D.L. (1977) A description of the NCAR global circulation models. in: *Methods in Comp. Physics Vol II, General Circulation Models of the Atmosphere*, (ed. J. Chang), Academic Press, New York, pp. 111–72.

Weinberg, A.M. (1985) Science and its limits: the regulator's dilemma. *Issues in Science and Technology*, **2**, 59–73.

Williams, J., Barry, R.G. and Washington, W.M. (1974) Simulation of the atmospheric circulation using the NCAR global circulation model with ice age boundary conditions. *J. Applied Met.* **11**, 305–17.

Wulff, F., Field, J.G. and Mann, K.H. (1989) *Network Analysis of Marine Ecosystems: Methods and Applications*. Coastal and Estuarine Studies Series, Springer-Verlag, Heidleberg.

4

Sustainable agriculture: the trade-offs with productivity, stability and equitability

Gordon R. Conway[1]

4.1 INTRODUCTION

The phrase 'sustainable agriculture' has acquired a diversity of meanings. To the agriculturalist, it means maintaining the momentum of the Green Revolution. To the ecologist it is a way of providing sufficient food without degrading natural resources. To the economist it represents an efficient long term use of resources, and to the sociologist and anthropologist it embodies an agriculture that preserves traditional values. Almost anything that is perceived as 'good' from the writer's perspective can fall under the umbrella of sustainable agriculture – organic farming, the small family farm, indigenous technical knowledge, biodiversity, integrated pest management, self-sufficiency, recycling and so on (Conway and Barbier, 1990).

This diversity of interpretation is to be welcomed as part of a process of gaining consensus for radical change. But it is confusing and results in concepts and definitions of little practical value. The often quoted definition of sustainable development proposed by the World Commission on Environment and Development (The Brundtland Report) – 'development that meets the needs of the present without compromising the ability of future

[1] This chapter is based on Conway, G.R. (in press) The sustainability of agricultural development: trade-offs with productivity, stability and equitability. *Journal of Farming Systems Research and Extension.*

Economics and Ecology: New frontiers and sustainable development.
Edited by Edward B. Barbier. Published in 1993 by Chapman & Hall, 2–6 Boundary Row, London SE1 8HN. ISBN 0 412 48180 4.

generations to meet their own needs' – is valuable as a policy statement but is too abstract for farmers, research scientists or extension workers trying to design new agricultural systems and develop new agricultural practices. For them a definition is needed that is scientific, is open to hypothesis testing and experimentation, and is practicable.

In this chapter I elaborate on a definition of sustainable agriculture that I and my colleagues have been using in recent years. The definition arises within a new paradigm for agricultural development that goes under the name of **Agroecosystem Analysis (AEA).** The origins of AEA lie in efforts to improve the analysis of natural ecosystems (Walker *et al.*, 1978) but most of the concepts and techniques were developed at the University of Chiang Mai in Thailand beginning in 1978 (Gypmantasiri *et al.*, 1980) and refined and elaborated at the University of Khon Kaen, also in Thailand, by a group of university and government research workers in Indonesia (KEPAS) and by the Southeast Asia Universities Agroecosystems Network (SUAN). The first major workshop on sustainable agriculture was held in Indonesia in 1982 (KEPAS, 1984).

The literature on AEA includes a number of books and papers (Conway, 1985, 1986, 1987; Conway and Barbier, 1990); most of the field based analyses are described in published reports (Conway and Sajise, 1986; Conway *et al.*, 1985, 1989; Ethiopian Red Cross, 1988; KEPAS, 1985a, b, 1986; KKU-Ford Cropping Systems Project, 1982a, b). Readers are also referred to the Gatekeeper Series on Sustainable Agriculture produced by the International Institute of Environment and Development in London.

4.2 AGROECOSYSTEMS

Agroecosystems are ecological systems transformed for the purpose of agriculture. For example, ricefields are created out of a swamp. Each field is formed by building up a bund that defines the biophysical boundary (Figure 4.1). Inside the boundary the great diversity of the original wildlife is reduced to a restricted assemblage of crops, pests and weeds – although still retaining some of the natural elements, such as fish and predatory birds. The basic, renewable ecological processes remain: competition between the rice and the weeds, herbivory of the rice by the pests, and predation of the pests by their natural enemies (and of the fish by the predatory birds). But these ecological processes are now overlain and regulated by the agricultural processes of cultivation, subsidy (with fertilizers), control (of water, pests and diseases), and harvesting. The result is an agricultural ecosystem or agroecosystem (Lowrance, Stinner and House, 1984; Spedding, 1975).

However, this is only a partial picture of the transformation. The agricultural processes are, in turn, regulated by economic and social decisions. Rice farmers cooperate or compete with one another and market, exchange

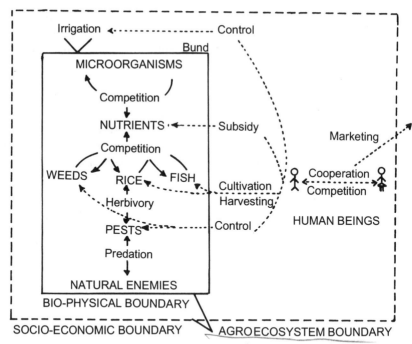

Figure 4.1 The ricefield as an agroecosystem. Solid lines represent ecological processes while dotted lines indicate the overlying agricultural and socio-economic processes. Source: Conway, 1987.

or consume their produce. The resulting system is as much a socio-economic system as it is an ecological system, and has a socio-economic boundary although it is not as easy to define as the biophysical one. This new, complex, agro-socio-economic–ecological system, bounded in several dimensions, I call an agroecosystem.

More formally, an agroecosystem is 'an ecological and socio-economic system, comprising domesticated plants and/or animals and the people who husband them, intended for the purpose of producing food, fibre or other agricultural products'.

A virtue of this definition is its emphasis on the interdisciplinary nature of agroecosystem structure and function. Despite a commitment to integrated analysis, biological and social scientists often work separately, at best coming together to write a final synthesis. Yet many, if not most, of the crucial questions for agricultural development lie not in one province or the other, but at their intersection. In Figure 4.1 the critical dynamics lie where socio-economic and agricultural processes intersect with the ecological, where fertilizer pricing meets plant competition or water supply affects weed control or pesticides interfere with natural enemies.

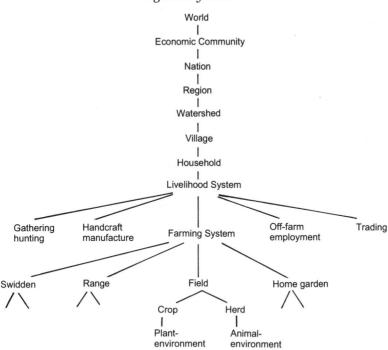

Figure 4.2 The hierarchy of agroecosystems. Source: Conway and Barbier, 1990.

4.2.1 The agroecosystem hierarchy

Agroecosystems defined in this way fall into a hierarchy. At the lowest level is the individual plant or animal, its immediate microenvironment and the people who tend and harvest it (Figure 4.2). The next level is the crop or herd, contained within a field or paddock, or in a swidden, home garden or range. These systems, alone or in various combinations, comprise a farming system. The hierarchy continues upwards in a similar fashion, each agroecosystem forming a component of the agroecosystem at the next level.

A feature of such hierarchies is that each level has a distinctive behaviour which is related to the behaviour of other levels, but not in a straightforward manner (Vickers, 1980). In practice, the properties of higher levels in such a hierarchy are not readily understood simply from a study of lower levels, and vice versa. As a consequence, each agroecosystem at each level of the hierarchy invites analysis in its own right.

4.2.2 Goals and strategies

Agroecosystems are cybernetic (Ashby, 1956). They have recognizable goals and strategies to attain them. I suggest that the primary goal of an agroecosystem is increased 'social value'. Broadly it is composed of the amounts of goods and services produced by an agroecosystem, the degree to which they satisfy human needs and their allocation among the human population. It also has a time dimension, since we seek not only increased benefits in the immediate future but also a degree of security over the longer term.

Thus social value has several measurable components: the present production of the agroecosystem, its likely level in the future and its distribution among the human population. These are expressed in four agroecosystem properties or behaviours (Figure 4.3):

1. productivity – the output of valued product per unit of resource input;
2. stability – the constancy of productivity in the face of small disturbing forces arising from the normal fluctuations and cycles in the surrounding environment;
3. sustainability – the ability of the agroecosystem to maintain productivity when subject to a major disturbing force;
4. equitability – the evenness of distribution of the productivity of the agroecosystem among the human beneficiaries, i.e. the level of equity that is generated.

4.3 THE MEASUREMENT OF AGROECOSYSTEM PROPERTIES

4.3.1 Productivity

The productivity of plants and animals can be measured as the amount of new biological material, or biomass, produced per unit of time. But agriculture is only concerned with the portion of the biomass that is useful – the harvest. Productivity is thus more appropriately measured either as yield or income or in terms of the other benefits that derive from the harvest.

Commonly, yield is measured per hectare or as the total production of agricultural goods and services per household, region or nation, but a large number of different measures are possible, depending on the nature of the product and of the resource input being considered. The yield may be in terms of kg of grain, tubers, meat or fish or any other consumable or marketable product. Alternatively, it may be expressed in calories, proteins or vitamins or as its monetary value at the market. In the last case it is measured as income over expenditure, i.e. as profit.

The three basic resource inputs to productivity are land, labour and capital. Strictly speaking energy is subsumed under land (solar energy), labour (human energy) and capital (fuel energy). Similarly technological inputs, such as fertilizers and pesticides, are components of capital, but both energy and technology can be treated for many purposes as separate inputs. Productivity can be tonnes of grain per hectare or per kg of nitrogen

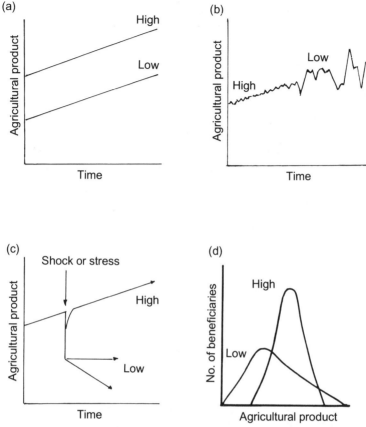

Figure 4.3 The properties of agroecosystems. Productivity measures valued (a) production, (b) stability (how it varies), (c) sustainability (how durable it is) and (d) equitability (how it is shared). Assessed together they indicate the present and expected social value of an agroecosystem. Source: Conway and Barbier, 1990.

fertilizer or per gram of active ingredient of insecticide, or any other conceivable combination of output and input.

4.3.2 Stability

Over time the productivity may rise or fall or remain constant. It will also exhibit a pattern of variability about the dominant trend line. The yield of a crop, for example, is likely to mirror the variability in the climate. Income may also fluctuate, not only reflecting changes in yield but also variations in the market price of inputs, such as labour, fertilizers and pesticides, and of the product. The latter, in turn, is a function of supply and demand.

Productivity may be defined in any of the ways described above and its stability measured by the **coefficient of variation in productivity**.

4.3.3 Sustainability

Stability measures the behaviour of an agroecosystem in response to the normal fluctuations in the surrounding environment. Productivity goes up and down but is not seriously threatened. However, agroecosystems are also subject to major disturbing forces which can cause productivity to fall well below its previous level. If productivity does fall it may recover either to its original level or to a new lower level or, in extreme situations, may cease altogether. Sustainability is the ability of an agroecosystem to withstand such disturbing forces.

The simplest case is an individual crop plant or animal. How does it withstand forces that threaten its survival? If we consider a plant strictly as a physical structure then the concepts of mechanics and Newton's laws of motion apply. A force acting on the plant, such as a low temperature, elicits a physical or chemical change. The force is termed a stress and the change a strain. If the change is reversible then the strain is regarded as elastic; if irreversible, the strain is plastic. When a maize plant is cooled from 30°C to 5°C it stops growing, but normal growth resumes when the temperature rises again. If wheat is cooled to 5°C growth continues, although at a slower rate. Both plants have suffered **elastic strain**, although the strain is greater in the case of maize. At an even lower temperature a plant dies, i.e. suffers a **plastic strain**.

But this physical model is too simple. The responses of biological systems are more than mechanical. Plants or animals may counter stress, either through evolutionary change or through adaptation during the organism's life. A common example of the latter is when plants of temperate regions harden to the freezing temperatures of winter through gradual exposure to the increasing cold of the autumn.

More dynamic still are the responses to non-physical forces, such as attacks by pests or competition from weeds. And the variety of forces and their responses becomes even greater as we move up the agroecosystem hierarchy through the crop, farm, village and nation. Productivity may collapse, for example, under the pressure of economic forces, such as a steep rise in input costs or growing indebtedness. Social disturbances, such as communal conflict or political revolution, may also pose a threat.

We can distinguish two kinds of disturbing force. First, there are forces that are relatively small and predictable, act on a regular and sometimes continuous basis and produce a large cumulative effect. Salinity, toxicity, erosion, pest or disease attack, indebtedness or declining market demand are examples. Such a force constitutes a **stress**. The other kind of force is one that is very large, infrequent and relatively unpredictable, and produces an immediate, large disturbance or perturbation. This I refer to as a **shock**. Examples are rare floods or droughts, or an outbreak of a new pest or the sudden rise in an input price. In practice, it is usually not difficult to distinguish between a stress and a shock, providing it is clear which level in the agroecosystem is being considered. A shock to an individual plant, say being destroyed by a pest, may be only part of a stress to the whole crop.

There is a continuum between stability and sustainability, but they are usually distinguishable by qualitatively different patterns of behaviour. Stability refers to the dynamics of an agroecosystem when subject to relatively minor and commonplace disturbing forces. Typically they are forces whose impact is small because, out of long association, agroecosystems have developed adequate defences. By contrast, sustainability is concerned with forces that are rarer, and less expected, so that agroecosystems are likely to have fewer or less well developed defences. Often the defence mechanisms are qualitatively different. An animal experiencing small changes in the outside temperature may respond with minor changes in circulation, but a sudden shock, such as immersion in freezing water, may elicit a qualitatively different response, for example the animal may secrete adrenaline.

Stability is easily measured from a time series of productivity. But the measurement of sustainability is more complex because of the range of forces and responses that may be encountered. The strength and nature of the shock or stress has to be assessed, the pattern of response to disturbance has to be described and the degree of resistance or resilience quantified. How far is the productivity depressed, how quickly and to what level does it return, and in what fashion?

Various responses are possible. First, productivity may remain unaffected. The agroecosystem resists the disturbing force and no disturbance occurs (Figure 4.4(a)). One mechanism of resistance is to **escape**. For example the seeds of wild cereals lie dormant in the ground throughout the Mediterranean summer, so avoiding the effects of the extreme heat. Nomadic pastoralists may move their cattle to escape an impending drought. Alternatively the agroecosystem may be **protected**. Farmers may build a bund to prevent flooding of crop fields. In analogous fashion a tariff wall, as has been erected by the European Economic Community (EEC), may protect crop production from falling world prices. A measure of resistance is **inertia** i.e. the size of the stress or shock that can be withstood without the productivity being affected. For example, where a bund is built to guard a crop the inertia can be expressed as the maximum height of flood against which the bund provides protection.

If resistance is weak then productivity may fall to below the usual range of variation. The question then is: how resilient is the agroecosystem? Will it return to the previous level or trend of productivity and how quickly and

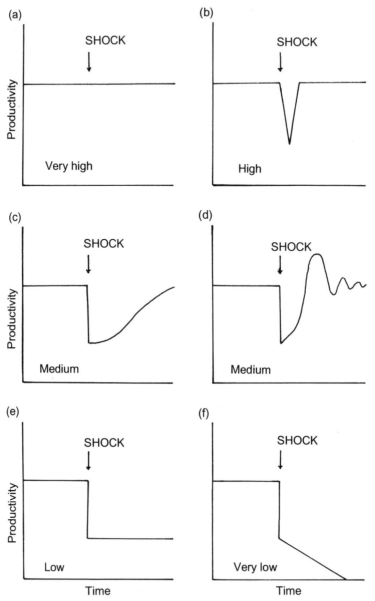

Figure 4.4 Patterns of response of productivity to shocks (The diagrams assume a level trend line, but this may be rising or falling.)

in what manner? The return may be smooth or the disturbance may continue with violent after-effects before the previous level or trend of productivity is regained (Figure 4.4(b)–(d).

Table 4.1 Measures of agroecosystem sustainability. All measures relate to the sustainability of the productivity trend. Based on Orians, 1975 and Westman, 1978

Characteristic	Definition	Measurement
Inertia	Resistance of productivity trend to change	Level of stress or shock that can be resisted without affecting trend line
Elasticity	Rapidity of recovery of productivity trend following disturbance	Time taken for trend line to be recovered
Amplitude	Zone from which recovery occurs following disturbance	Maximum amount of disturbance following stress and shock from which recovery is possible
Hysteresis	Degree to which path of recovery is exact reversal of disturbance path	Difference between disturbance and recovery paths
Malleability	Degree of difference between system state before and after disturbance	Difference between mean productivities

Sometimes the productivity may not recover: it may remain at a new lower level or trend line, or in extreme situations continue to fall and disappear altogether (Figure 4.4(e)–(f)). Various measures of this pattern of resilience are possible (Table 4.1). **Elasticity, amplitude** and **hysteresis** measure the speed, likelihood and pattern of recovery of productivity following a stress or shock. **Malleability** measures the difference between the mean productivity before and after the disturbance.

Studies of the resilience of ecological systems gained widespread attention following the work of C.S. Holling (1985). He stressed the importance of distinguishing, as I do here, between the behaviour of systems under normal environmental conditions, i.e. their stability, and when subject to a disturbance which can potentially dramatically change their state. He suggested that ecological and other systems could potentially exist in more than one steady state and when subject to disturbance, could 'flip' from one state to another. There is some dispute as to whether multiple steady states exist in natural ecological systems (Connell and Sousa, 1983), but they commonly occur in agroecosystems. An example is the collapse of swidden cultivation from a regular cropping/fallow cycle to a persistent *Imperata* grassland cover following the stress of increased crop harvesting (Figure 4.5).

Another example is the behaviour of grazing systems, investigated by Noy-Meir (1975). Here increasing livestock raises productivity but also stresses the vegetation. At a certain intensity of stress, which is often very close to the maximum livestock carrying capacity, the vegetation collapses

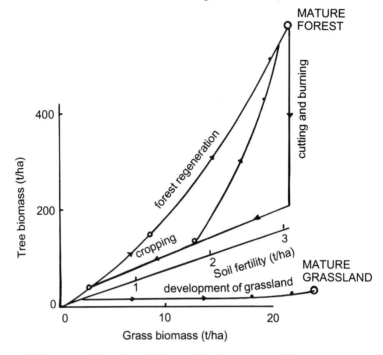

Figure 4.5 Sustainability of a shifting cultivation system. When up to eight crops are grown in the cropping phase, the system goes back to mature forest but after eight crops or more it moves toward permanent grassland. Eight crops is thus a measure of the amplitude. Source: After Trenbath, Conway and Craig, 1990.

and the grazing system moves to a new level of productivity, much lower than before.

Finally, an assessment of sustainability requires a measure of the effectiveness of internal adjustments that agroecosystems make in response to stresses and shocks. The variety of such adaptations is considerable. Hardening is one example. Another is the construction of a bund to protect against flooding. Weeds may be countered by hoeing or by more vigorous crop growth. At the farm level the response to growing debt may be to switch to a less risky crop/livestock combination or to one requiring lower inputs. A village stressed by the loss of young people emigrating in search of more lucrative work may respond by adopting more labour saving techniques of cultivation. Similarly, a district may counter rising transport costs by a switch to higher value, lower volume products; a region may respond to a widespread drought by establishing a network of famine relief stores as a protection against future droughts; and a nation may respond to increasing competition by changing the nature of its productivity so as to exploit its comparative advantage with respect to other nations.

4.3.4 Equitability

Productivity, stability and sustainability adequately measure how much an agroecosystem produces and is likely to produce over time. An African village agroecosystem that produces a high, stable yield of sorghum using practices and varieties that are broadly resistant to pests and diseases will have a higher social value than another village producing lower, less stable and sustainable yields. However, social value depends not only on the pattern of production but also on the pattern of consumption. Who benefits from the high, stable and sustainable production? How is the harvested sorghum, or the income from the sorghum, distributed among the people of the village? Is it evenly shared or do some villagers benefit more than others?

Equitability describes this pattern of distribution of productivity. Again, productivity may be measured in any of the ways described above. Thus the product per man hour of a farm may be shared between the tenant and the landowner. More commonly, equitability refers to the distribution of the overall production, that is the total goods and services produced by an agroecosystem. In subsistence farming the producers are all consumers, but the higher the agroecosystem in the hierarchy and the greater the degree of commercialization, the more do non-producers benefit.

The most straightforward way of measuring how evenly products or income are distributed is to construct a histogram. But this can give a misleading impression since histograms tend to overemphasize the middle of the distribution at the expense of the extremes (Atkinson, 1970). At the left, beyond the origin, are those with negative incomes – bankrupt farmers or those close to starvation. While to the right the histogram has to be impossibly extended to incorporate the rich.

Alternatively there are a number of other ways of representing equitability that are intended to be positive, i.e. objective measures of distribution. The most commonly used is the Lorenz curve and the Gini coefficient derived from it (Gini, 1912; Sen, 1973; Lorenz, 1905; Atkinson, 1970). However, these measures suffer from several drawbacks (see the review in Sen, 1973). For example, they tend to give relatively greater weight to changes in different parts of the range. A more important criticism is that they refer to distribution of income or product among people as though everyone is alike. This is clearly untrue. People differ in their endowments and derive different amounts of individual utility from their income. A kilogram of rice is valued much more by a person who is starving than by one who is rich. Furthermore, individuals compare what they receive with what others receive. Measures of equitability, if they are to reflect the actual decisions that social groups (whether they be households, villages or nations) make, must incorporate these notions of social justice. But in practice this is hard to accomplish since measurement of individuals' needs and utilities are difficult and laborious. The most satisfactory measure, so far, is that of Atkinson (1975) who weights the individual incomes before adding them.

4.4 TRADE-OFFS

Each agroecosystem, at each level in the hierarchy, has a social value. And one kind of agroecosystem may have a greater social value than another and hence is more sought after. I assume that people seek to maximize social value and, to this end, will adopt strategies consisting of various combinations of productivity, stability, sustainability and equitability. As in the case of individual organisms, there are inevitable trade-offs between the levels of the properties. For example, a large-scale irrigation project may achieve greater overall productivity yet be at the expense of sustainability and equitability. Similarly, too much emphasis on equitability may inhibit productivity.

4.4.1 Quantitative analysis

To date there has been very little quantitative analysis of the trade-offs between agroecosystem properties. A notable example is the analysis carried out by Pretty (1990) of the properties of manorial agriculture in England during the Middle Ages. The study was based on a remarkable set of records for individual manors extending from AD 1283 to 1349 (Titow, 1972).

Three main cereals were grown – oats, wheat and barley – and their productivity, stability and sustainability are summarized in Table 4.2. Productivity was measured as the number of seeds harvested per seed sown or as net yield per hectare (gross yield minus seed retained for the next sowing). On both measures, wheat and barley outyielded oats but the productivity of oats was notably more stable. Pretty (1990) has also shown that the price of oats was more stable than that of wheat and barley.

Table 4.2 Yields of cereals grown on English manors from 1283–1349. Source: Pretty, 1990

	Wheat	Oats	Barley
Productivity			
Net yield (kg/ha)	385	300	540
Seeds/seed sown	4.0	2.3	3.5
Stability (coefficient of variation)			
Net yield	38.8%	31.3%	39.9%
Seeds/seed sown	36.9%	33.6%	37.3%
Sustainability (elasticity)			
Time to return to average	4–7+ yrs	1–5 yrs	-

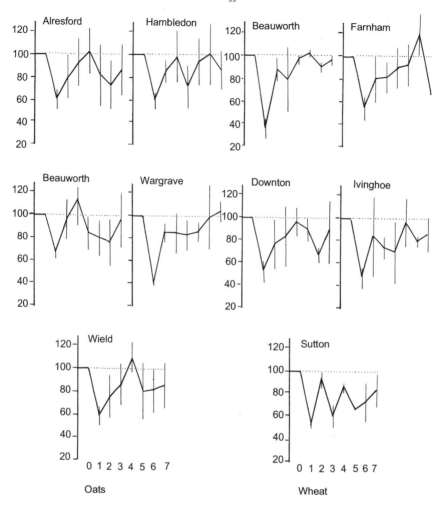

Figure 4.6 Responses of oat and wheat yields following a major disturbance to productivity. Source: Pretty, 1990.

Sustainability of the cereals was assessed by plotting the pattern of yields following a significant fall in yield, caused by a very wet year for example. Figure 4.6 shows the contrasting patterns of response for wheat and oats. After poor harvests oats recover more rapidly than wheat, i.e. are more **elastic** (Table 4.2). Its greater stability and sustainability meant that the oat crop was more reliable and hence often provided the mainstay of the peasant's diet, particularly in the more marginal lands.

The manorial agroecosystem of England was extremely long-lived and Pretty (1990) has concluded that the high degree of sustainability was marked by relatively high stability and equitability, but these were obtained at the expense of productivity. Yields had not significantly increased since Roman times and remained low for virtually all of the manorial period.

4.4.2 Qualitative analysis

Most analyses of the trade-offs between agroecosystem properties are qualitative. They typically consist of rapid assessments of the likely impact of innovations proposed during AEA exercises.

An example is the AEA carried out in the Wollo province of Ethiopia in 1986 (Ethiopian Red Cross, 1988). The exercise lasted just under two weeks and was carried out by a multidisciplinary team comprising headquarters staff of the Ethiopian Red Cross, the Ministry of Agriculture and field staff from the province. Two villages were analysed using RRA techniques with an emphasis on semi-structured interviewing, analytical games such as preference ranking, and on constructing maps, transects and seasonal calenders. From these problems and opportunities were identified, and a shortlist produced of innovations or 'best bets' that the team felt would be appropriate for development of the two villages. These best bets were then prioritized in terms of their cost and feasibility and their likely impact on development in terms of the system properties (Table 4.3).

The best bets had been chosen for their potentially high stability and sustainability but analysis by the team suggested considerable differences in potential productivity and equitability. Two of the best bets with the highest productive potential – lowland irrigation and the introduction of rainfed crops – were judged to have little effect in improving equitability, primarily because they would only benefit certain better-off farmers. Other bets with a high potential for benefiting poor farmers were considered relatively less productive. The innovation which scored consistently high across the board was the development of home gardens for the new village sites that the government was in the process of creating. One outcome of the exercise was the establishment, by the Ethiopian Red Cross in the weeks that followed, of a nursery of home garden plants.

Because of the short time of the exercise and the pressing need to decide on appropriate innovations for the impoverished villages, no detailed quantitative analysis was possible. Assessment was based on the collective judgement of the team members, based on their experience and intuition. Although the results lacked precision or rigour, they had the advantage of being collectively agreed upon and this facilitated the subsequent action that was taken.

4.4.3 Minimizing the trade-offs

The social value of an agroecosystem is the product of the levels of the four

Table 4.3 Assessment of best bets for Abicho, Wollo province, Ethiopia.
Source: Ethiopian Red Cross, 1988

Bet/Innovation	Productivity	Stability	Sustainability	Equitability	Cost	Feasibility	
						Technical	Social
1. Lowland irrigation	++	+	++	0	X	XX	XXX
2. Gully cropping	+	+	++	++	XX	XXX	XXX
3. New rainfed crops	++	++	++	0	XX	XX	XX
4. Upland revegetation	+	++	++	++	X	XXX	X
5. New forage crops	+	++	++	+	XX	XX	XX
6. Household water supply	+	++	++	+	XX	XX	XXX
7. Home garden development	++	++	++	++	XXX	XXX	XXX

-	negative impact	X	high cost or poor feasibility
0	no impact	XX	medium cost or feasibility
+	positive impact	XXX	low cost or high feasibility
++	very positive impact		

different system properties. However the product is not a simple arithmetic addition or multiplication, since people in each place and at each period of time will differently weight the properties (Conway, 1985). In the Middle Ages high stability and sustainability appear to have been given a preference over high productivity. In the Ethiopian AEA the team engaged in a fierce dialogue over the relative weighting between productivity and equitability when deciding on their final priorities. Nevertheless, in most situations, there is a preference for agroecosystems in which all properties are high and the trade-offs are explicitly minimized. In the Ethiopian case the most promising innovation in this respect was home garden development.

Home gardens are one of the oldest forms of farming system and may have been the first agricultural system to emerge in hunting and gathering societies. Today, home or kitchen gardens are particularly well developed in the island of Java in Indonesia and these have been explicitly analysed in terms of agroecosystem properties by Soemarwoto and Conway (in press).

The immediately notable characteristic of home gardens is their great diversity relative to their size. In one Javanese home garden 56 different species of useful plants were found; also commonly present is a diversity of livestock – cattle, goats, chickens, fish in fish ponds and so on. Closer analysis shows the high diversity to be matched by high levels of productivity, stability, sustainability and equitability (Table 4.4).

Table 4.4 System properties of the home garden when compared with a rice field. Source: Soemarwoto and Conway, in press

	Home garden	Rice field
Productivity	Higher standing biomass Higher net income (lower inputs) Greater variety of production (food, medicines, fuelwood)	Higher gross income
Stability	Year round production ('living granary') Higher year to year stability	Seasonal production Vulnerable to climatic and disease variation
Sustainability	Maintenance of social fertility Protection from soil erosion	Heavy pest and disease attack
Equitability	Home gardens in most households Barter of products	Product to landowners

Part of the reason for this minimal trade-off is the inherent diversity. It helps stabilize production, buffers against stress and shock and contributes to a more valued level of production. But equally important is the intimate nature of the home garden. The close attention that is possible from family labour ensures a high degree of stability and sustainability and the link between the garden and the traditional culture leads to an equitable distribution of the diverse products.

4.5 CONCLUSIONS

In many respects the home garden is a unique agroecosystem. But undoubtedly there are other systems that have similarly high levels of the four properties and, more important, there are technologies and innovations which can potentially help reduce the trade-offs between the properties. The methods of analysis described above provide a rigorous yet straightforward approach to assessing agroecosystem properties and their trade-offs. Experience in an increasing number of field projects has demonstrated that it is an approach that is practical and easy to use in the hands of people with a range of disciplinary skills and backgrounds. As the Ethiopian case study demonstrates, it can be easily incorporated in day to day decision making.

Defining sustainability in terms of preservation or duration, as is commonly done, has little practical value. Long-term experiments to measure persistence, for example of different cropping systems, are of research interest but take too long to constitute a practicable analytical method. By contrast, measuring the ability of an agroecosystem to withstand stress and shock is a subject for experiments using classical agricultural methods.

The current danger of using sustainability as a loosely defined term to encompass a wide range of systems and technologies is that benefits may be obtained at the expense of other, less obvious costs. High sustainability is not the only desirable aspect of agricultural production and in many situations it may be necessary to trade a degree of sustainability for higher levels of productivity or equitability. Choices and decisions are necessary and are best made when the options are clearly apparent.

REFERENCES

Ashby, W.R. (1956) *An Introduction to Cybernetics*, Chapman & Hall, London.
Atkinson, A.B. (1970) On the measurement of inequality. *Journal of Economic Theory*, **2**, 244–63.
Atkinson, A.B. (1975) *The Economics of Inequality*, Clarendon Press, Oxford.
Connell, J.H. and Sousa, W.P. (1983) On the evidence needed to judge ecological stability or persistence. *The American Naturalist*, **121**, 789–824
Conway, G.R. (1985) Agroecosystem analysis. *Agricultural Administration*, **20**, 31–55.

Conway, G.R. (1986) *Agroecosystem Analysis for Research and Development*, Winrock International, Bangkok.

Conway, G.R. (1987) The properties of agroecosystems. *Agricultural Administration*, **24**, 95–117

Conway, G.R. and Barbier, E.B. (1990) *After the Green Revolution: Sustainable Agriculture for Development*, Earthscan, London.

Conway, G.R. and Sajise, P.E. (1986) *The Agroecosystems of Buhi: Problems and Opportunities*, Los Banos, Program on Environmental Science and Management, University of the Philippines.

Conway, G.R., Alam, Z., Husain, T. and Mian, M.A. (1985) *An Agroecosystem Analysis for the Northern Areas of Pakistan*, Gilgit, Pakistan, Aga Khan Rural Support Programme.

Conway, G.R., Sajise, P.E. and Knowland, W. (1989) Lake Buhi: resolving conflicts in a Philippine development project. *Ambio*, **18**, 128–35.

Ethiopian Red Cross Society (1988) *Rapid Rural Appraisal: A Closer Look at Rural Life in Wollo*, Addis Ababa, Ethiopian Red Cross Society, London, International Institute for Environment and Development.

Gini, C. (1912) *Variabilita e Mutabilita*, Bologna.

Gypmantasiri, P., Wiboonpongse, A., Rerkasem, B., *et al.* (1980) *An Inter-disciplinary Perspective of Cropping Systems in the Chiang Mai Valley: Key Questions for Research*, Chiang Mai, Thailand, Faculty of Agriculture, University of Chiang Mai.

Holling, C.S. (1985) Perceiving and managing the complexity of ecological systems, in *The Science and Praxis of Complexity*, United Nations University, Tokyo.

KEPAS (1984) *The Sustainability of Agricultural Intensification in Indonesia: A Report of Two Workshops of the Research Group on Agroecosystems*, Jakarta, Indonesia: Agency for Agricultural Research and Development.

KEPAS (1985a) *The Critical Uplands of Eastern Java: An Agroecosystem Analysis*, Jakarta, Indonesia, Agency for Agricultural Research and Development.

KEPAS (1985b) *Swampland Agroecosystems of Southern Kalimantan*, Jakarta, Indonesia, Agency for Agricultural Research and Development.

KEPAS (1986) *Agro-ekosistem Daerah Kering di Nusa Tenggara Timur*, Jakarta, Indonesia, Agency for Agricultural Research and Development.

KKU-Ford Cropping Systems Project (1982a) *An Agroecosystem Analysis of Northeast Thailand*, Khon Kaen, Thailand, Faculty of Agriculture, Khon Kaen University.

KKU-Ford Cropping Systems Project (1982b) *Tambon and Village Agricultural Systems in Northeast Thailand*, Khon Kaen, Thailand, Faculty of Agriculture, Khon Kaen University.

Lorenz, M.O. (1905) Methods of measuring the concentration of wealth. *Journal of the American Statistical Association* **9**, 209–19.

Lowrance, R., Stinner, B.R. and House, G.J. (eds) (1984) *Agricultural Eco-systems: Unifying Concepts*, John Wiley, New York.

Noy-Meir, I. (1975) Stability of grazing systems: an application of predator–prey graphs. *Journal of Ecology*, **63**, 459–81

Orians, G.H. (1975) Diversity, stability and maturity in natural ecosystems, in *Unifying Concepts in Ecology* (eds W.H. van Dobben. and R.H. Lowe-McConnel), Junk, The Hague, pp. 64–5.

Pretty, J.N. (1990) Sustainable agriculture in the Middle Ages: The English manor. *The Agricultural History Review*, **38**, 1–19.

Sen, A. (1973) *On Economic Inequality*, W.W. Norton, New York.

Soemarwoto, O. and Conway, G.R. (in press). The Javanese home garden. *Journal of Farming Systems Research and Extension*, **2**, 95–117.

Spedding, C.R.W. (1975) *The Biology of Agricultural Systems*, Academic Press, London.

Spedding, C.R.W. (1979) *An Introduction to Agricultural Systems*, Applied Science Publishers, London.

Titow, J.Z. (1972) *Winchester Yields*, Cambridge.

Trenbath, B.R., Conway, G.R. and Craig, I.A. (1990) Threats to sustainability in intensified agricultural systems: analysis and implications for management, in *Agroecology: Researching the Ecological Basis for Sustainable Agriculture*, (ed S.R. Gleissman), Springer-Verlag, New York , pp. 337–65.

Vickers, G. (1980) *Responsibility – Its Sources and Limits*, Intersystems Publications, Seaside, California.

Walker, B.H., Norton, G.A., Conway, G.R. *et al.* (1978) A procedure for multidisciplinary ecosystem research: with reference to the South African Savanna Ecosystem Project. *Journal of Applied Ecology*, **15**, 481–502.

Westman, W.E. (1978) Measuring the inertia and resilience of ecosystems. *Bio Science*, **28**, 705–10.

Wiener, N. (1948) *Cybernetics*, MIT Press, Cambridge and John Wiley, New York.

5

Stress, shock and the sustainability of optimal resource utilization in a stochastic environment

Charles Perrings

5.1 INTRODUCTION

The close links between agricultural and economic growth in the low income countries of sub-Saharan Africa during a decade of 'crisis' has prompted a widespread theoretical and empirical reappraisal of the dynamics of agricultural sector performance in such countries. A common thread running through this reappraisal is the observation that falling agricultural productivity reflects the widespread degradation of the resource base. As Pearce, Barbier and Markandya (1988) put it: 'Africa's economic crisis certainly appears to be largely due to Africa's agricultural crisis', and 'Africa's agricultural crisis is in significant part an environmental crisis'. A variety of explanations for the trends that underlie this 'crisis' are offered in the literature. The dominant view, however, is that the causes of an ecologically unsustainable use of agricultural resources are to be found in the economic environment within which farmers make their decisions (Repetto, 1986, 1989; Warford, 1989; Lutz and El Serafy, 1989; Barbier, 1989b).

However, while we may agree that the creation of an 'appropriate economic environment' is a precondition for the sustainable use of natural resources in agriculture, it is not at all clear what the term implies under

Economics and Ecology: New frontiers and sustainable development.
Edited by Edward B. Barbier. Published in 1993 by Chapman & Hall, 2–6 Boundary Row, London SE1 8HN. ISBN 0 412 48180 4.

the ecological conditions prevailing in sub-Saharan Africa – or anywhere else for that matter. There is a very strong tendency to equate 'an appropriate economic environment' with the liberalization of agricultural product and financial markets. Indeed, liberalization of agricultural prices in the low income countries has been widely endorsed on the grounds that it implies higher producer prices, and it is held that these will provide both the means and the incentive to conserve the agricultural resource base (Bond, 1983; Cleaver, 1985, 1988; Barbier, 1988, 1989b). But this endorsement is not founded on any clearly understood relation between the economic and ecological components of agricultural systems in the region. Whether any bioeconomic system is sustainable depends on both the ecological and the economic parameters of that system, and there is no reason to believe, a priori, that world prices are more compatible with the sustainability of the system than any other prices. The case for the liberalization of agricultural output prices in the low income countries may be strong, but it does not rest on an environmental foundation.

At the most general level, this paper is concerned with the constraints imposed by the economic environment on the sustainable use of ecological resources. To focus discussion, I consider the case of pastoralism in semi-arid areas such as the Sahel and much of eastern and southern Africa. The principal management problem in this case is to determine the optimal level of activity over time, given the evolution of the highly uncertain natural and economic environments within which pastoralists operate. Pastoralism may be said to be sustainable if the solution to this management problem does not involve the degradation of the natural resources required by the activity. Sustainability refers to the resilience of an ecological system subject to stress (determined by the level of economic activity) in the face of shocks (determined by climatic variation). This interpretation of sustainability, due to Conway and Barbier (Conway, 1987; Conway and Barbier, 1990; and Barbier, 1989a), directs our attention to the relation between the economic environment as the prime determinant of activity levels, and a stochastic ecological environment as the source of shocks.

The value of deterministic bioeconomic control models in the management of ecological resources in agriculture in the low income countries has recently been stressed by Barrett (1989a, 1989b). Since an essential feature of rain-fed agriculture in semi-arid areas is that it is subject to a high degree of uncertainty related to the variance in rainfall, however, the management problem in this case is treated stochastically. More particularly, it is treated as an infinite horizon stochastic control problem, in which the state variables are the size of the herd and the carrying capacity of the range, and the control variable is the level of offtake. A stochastic control formulation turns out to be especially useful in analysing the long term effects of a given price structure under conditions of ecological uncertainty. While there is reason to believe that it is the long term effects of current price regimes that

se for alarm in the semi-arid lands, there are few attempts to
_ate those long term effects in an analysis of price policy. If it
_rectly describes the processes involved, the stochastic control model can
identify the effect of a given price structure on the sustainability of the time
paths of the state variables.

The chapter is organized in seven sections. The following section discusses the ecological component of the model, and clarifies the assumptions
made about range and herd dynamics in the absence of control. Section 5.3
elaborates the management problem addressed, and derives the equations
required for the construction of an optimal policy. In section 5.4 a simulation is offered that permits description of the properties of an optimal policy
in the stochastic case, and the properties of the time paths of both state and
control variables under given values for the ecological and economic
parameters. Sections 5.5 and 5.6 consider the impact of the economic
environment within which the problem is optimized, and a final section
offers some concluding remarks.

5.2 AN ECOLOGICAL MODEL

Three characteristics of the pastoral economy are assumed to be essential
to the specification of the ecological component of a bioeconomic model of
rangeland use:

1. Current changes in herd size and the level of offtake from herds affect
   future rangeland carrying capacity. This means both that the current
   carrying capacity of rangeland is not independent of the history of
   rangeland use, and that current herd size is not independent of the past
   evolution of carrying capacity. It is this interdependence of population
   and carrying capacity which makes the problem somewhat different
   from the many other renewable resource problems analysed in control
   terms.
2. The system evolves through periodic change. The 'process noise' that
   randomizes the time paths of herd size and carrying capacity is the
   product of variance in rainfall. Moreover, the effect of variance in rainfall
   on herd size and carrying capacity is not instantaneous, but occurs with
   a seasonally determined lag. Accordingly, time is treated discretely,
   rather than continuously, with a year being the natural interval. This
   assumption has certain fairly obvious implications for the time-behavi-
   our of the system, given that difference and differential equations do not
   behave in the same way.
3. There is a very large measure of uncertainty attached to the future value
   of the state variables – herd size and carrying capacity. As has already
   been remarked, the carrying capacity of rangeland and the growth rate
   of the herd are both a function of rainfall, which has an extremely high

variance in semi-arid areas. This characteristic makes the problem an intrinsically stochastic one.

To reflect these characteristics, the structure of the model developed here differs in certain respects from existing pastoral models. Like Barrett's (1989a) adaptation of the biological predator–prey models of May (1974), it postulates logistic growth functions for the herd and the vegetative cover of the range (which determines its carrying capacity), but as will become clear, treatment of the dynamics of carrying capacity is somewhat different.[1] Carrying capacity is assumed to change over time with changes both in the degree of grazing pressure, and in climatic conditions. The relation between carrying capacity and grazing pressure is defined within the model. The relation between carrying capacity and factors exogenous to the model, such as rainfall, is captured by random variation of the ecological parameters of the model. Just how the regeneration of the range is expected to change with variation in endogenous factors will be discussed momentarily, but the important point is that the ecological model should reflect the fact that herd growth, the impact of herd size on rangeland vegetation, and the recuperative powers of the range are interdependent, and are all sensitive to climatic conditions.

The ecological system in isolation is assumed to be globally stable, in that the vegetative cover of the range is assumed to converge to some well-defined maximum value (the climax vegation) regardless of the severity of shocks due to either temporary climatic variation or overgrazing. This reflects the 'resilience' hypothesis (Walker and Noy-Meir, 1982), which rules out the possibility that there is some threshold level in the quality of the resource below which it will collapse completely. If the ecological system is globally stable, then it can withstand any level of overgrazing, regenerating in the same way irrespective of the damage inflicted on both the soil and its vegetative cover. Put another way, the assumption implies that nothing is irreversible in the ecological system: it is non-evolutionary.

[1] Barrett's (1989a) model has the following structure. The growth equation for the herd is

$$H_t = rH_t(1 - H_t/K_t) - h_t$$

(1)

in which H_t denotes herd size at time t, r denotes the rate of growth of the herd, K_t denotes the carrying capacity of the range, and h_t denotes offtake at time t. The equation of motion for K_t is given by

$$K_t = a(\underline{K} - K_t) - bH_t$$

(2)

in which $\underline{K}$ is defined as the 'saturation level of the grazing lands', a denotes the natural rate of regeneration of the rangeland, and b denotes the rate of depletion by the herd.

It is assumed that the unit price of the offtake, p (>0), and the unit cost of maintaining the herd, c (>0), are both constant, implying that profits are given by $ph_t - cH_t$ for all t. The economic control problem is then to

$$\max_{(h_t)} \int_0^\infty [ph_t - cH_t]e^{-\delta t}dt \qquad (3)$$

s.t. (1) and (2), and with $H_0, K_0 > 0$; $H_t, K_t \geq 0$; and $0 \pounds h_t \pounds h^{max}$. The last restriction implies that there is some maximum level of offtake which is different from the size of the herd itself, and

ɔn in vegetative cover in the short run does not affect the climax
..ɔn in the long run this assumption is reasonable, but note that even
.ıt the short run it is not undisputed (Westoby, 1979). It is worth emphasizing that the assumption that the ecological system is globally stable does
not imply that the pastoral economy is globally stable. Indeed, the collapse
of carrying capacity and the extinction of the herd are possible outcomes
of an optimal strategy.

Leaving the evolutionary nature of the ecological system to one side, and
abstracting from the significance of herd and range composition, the remaining characteristics of pastoral systems seem to be adequately captured
in the following equations of motion. The sequences $\{x_t\}$ and $\{k_t\}$ describe
the time paths of the herd and the carrying capacity of the range. Both are
measured in terms of livestock units, and both may be thought of as the
natural endogenous variables of the system. $\{x_t\}$ and $\{k_t\}$ are generated by
the following first order forward recursions:

$$x_{t+1} - x_t = \alpha_t (1 - x_t/k_t) - u_t \tag{1}$$

$$k_{t+1} - k_t = \beta_t K_t(1 - k_t/k_c) - \gamma_t (x_t - u_t) \tag{2}$$

in which
x_t = herd size at time t ($0 \leq x_t \leq k_c$);
k_t = carrying capacity at time t ($0 \leq k_t \leq k_c$);
k_c = maximum carrying capacity of the range;
u_t = offtake at time t ($-k^c \leq u_t \leq x_t$);
α_t = the net growth rate of the herd on the range whose use is being
evaluated ($-1 \leq \alpha_t$);
β_t = the rate of regeneration of the range ($-1 \leq \beta_t$);
γ_t = the rate of depletion of the range due to the herd ($\gamma_t \leq 1$).
Equation (1) describes the net growth of the herd in any given period as the
difference between offtake and the natural growth of the herd, given the
degree of grazing pressure, x_t/k_t. Equation (2) describes the net growth in
the carrying capacity of the range as the difference between the net depletion of the vegetative cover of the range due to herd management policy,
and the natural rate of regeneration of the range. Offtake, u_t, accordingly
has both direct and indirect effects on the size of the herd. If livestock are
drawn off in the current period, the size of the herd is reduced. But, at the
same time, the future growth potential of the herd is improved due to the
reduction in the net depletion of the vegetative cover of the range. Much of
the interesting dynamics of the model are due to these indirect effects.

To obtain the maximum sustainable yield of the range given these
equations of motion, notice first that the maximum carrying capacity of the
range, k_c, is defined as the current carrying capacity of the climax vegetation
on virgin range. At that point the growth in carrying capacity or the rate of
regeneration of the range is zero, and the addition of any livestock will

result in a fall in carrying capacity.[2] Maximization of the growth function (2) with respect to current carrying capacity shows that the maximum rate of regeneration of the range occurs when current carrying capacity is one half of the maximum carrying capacity: that is, when $k_t = k_m = 1/2k_c$. The maximum sustainable yield of the range is the point at which the net rate of depletion of the range is equal to the maximum rate of its regeneration. From (1) and (2), the size of the herd corresponding to the maximum sustainable yield is given by:

$$x_m = (k_c/4\alpha_t) \ \{-(1 - \alpha_t) \pm [(1 - \alpha_t)^2 + 2\alpha_t \ \beta_t \ /\gamma_t \ ]^{\frac{1}{2}} \} \qquad (3)$$

and the maximum sustainable level of offtake is given by

$$u_m = \alpha_t \ x_m(1 - x_m/k_m) = x_m - 1/2(\beta_t \ /\alpha_t)k_m \qquad (4)$$

Whereas the state variables, x_t and k_t are restricted to non-negative values, offtake, u_t, may in principle be positive or negative. If offtake is positive (implying that livestock is being drawn off the range) it is limited to values less than or equal to the size of the herd. If offtake is negative (implying that the range is being restocked) it is limited to values less than or equal to the maximum carrying capacity of the range. Although a policy requiring negative offtake in the long run would not be economically interesting, restocking may be a part of an optimal strategy in the stochastic case.

The time behaviour of the ecological system depends on the ecological parameters, α_t β_t and γ_t. Note that the time behaviour of non-linear difference equations similar to (1) and (2) tends to be rather complex. Ignoring offtake, if α_t and β_t were assumed to be positive constant parameters (as they would be in the deterministic case), the size of the herd and the carrying capacity of the range would converge to equilibrium values so long as $0 < \alpha_t \ \beta_t \leq 2$. Moreover, convergence would be asymptotic for $0 < \alpha_t \ \beta_t \leq 1$, and through damped oscillation for $1 < \alpha_t \ \beta_t \leq 2$. But for $\alpha_t \ \beta_t \geq 2$ the sequences $\{x_t\}$ and $\{k_t\}$ would be non-convergent, and if $\alpha_t \ \beta_t > 2.57$, would exhibit 'chaotic' behaviour. In the general case where α_t and β_t are not restricted to positive values, and are time-varying, the properties of the system will change depending on the current value of α_t. $\{x_t\}$ and $\{k_t\}$ may converge on some positive growth path over some time segments, may converge on zero over others, or may be entirely non-convergent.

In this chapter, α_t β_t and γ_t are defined as stochastic parameters. That is, they are independently distributed random numbers with means $\underline{\alpha}, \underline{\beta}$ and $\underline{\gamma}$, and variances $\sigma\alpha^2, \sigma\beta^2$ and $\sigma\gamma^2$. The system is thus subject to 'process noise'.[3] To get a sense of the likely values for the mean and variance of α_t,

---

[2] This abstracts from the potential for increasing carrying capacity through bush clearance – argued to be important in some areas.

[3] In reality grazing systems are usually also subject to 'observation noise'. That is, the state variables x_t and y_t will tend to be observed only with some error. However, we may ignore this additional source of uncertainty here.

we need to consider both the ecological and institutional determinants of herd growth. Recall that the focus of the chapter is rangeland in semi-arid areas in the low income countries of sub-Saharan Africa. Institutionally, these areas tend to be dominated by more or less regulated common property regimes. While the main source of variance in α_t is rainfall (which affects the net natural rate of increase of the herd), it also depends on the incidence of disease (which need bear no particular relation to the level of rainfall).[2] The net natural rate of increase is a function both of fertility and mortality in the herd, each of which tends to be sensitive to the level of rainfall for most species herded in these areas. Pastoralism in most semi-arid areas involves the managed movement of herds between ranges, depending on the state of the vegetative cover – the pattern being facilitated by open access common property regimes. But there is also some autonomous movement both of livestock and of competing ungulates. This implies that while α will tend to have a value somewhere near the long-run mean net natural rate of increase of the herd itself, it will be subject to considerable variance. Indeed, given the influence that rainfall has on herd fertility, mortality and migration, fluctuation in rainfall has historically led to dramatic swings in herd sizes on a given range from one period to the next. It has also led to dramatic shifts in the species composition of the herd, although no attempt is being made to model this here.

The mean and variance of β_t capture the net natural rate of increase of the vegetation consumed by the herd on the rangeland of interest. As in the case of herd growth, there are a number of different effects involved here. Clearly, there is a positive correlation between rainfall and the growth of graze or browse, but where there is a short run shift in species composition within the vegetative cover from edible grasses to woody biomass, or from edible to non-edible grasses, this will show up as a decrease in vegetative cover, even though total biomass may have increased. Similarly, where climatic conditions associated with increasing vegetative biomass also favour the growth of populations of competitors to the herd (other ungulates or insects, say), the graze or browse available to the herd may decrease. This is a very real problem in many of the semi-arid areas of sub-Saharan Africa, which are also populated by highly mobile, highly fluctuating herds of antelope, and are subject to depredation by insect swarms. An additional complication arises if edible grasses are not the climax vegetation of the area, but this problem is set aside in this chapter. Once again, while β will tend to have a value somewhere near the long-run, mean, net natural rate of regeneration of the range, it will be subject to considerable variance.

As a result of the variance of α_t and β_t, the time behaviour of the recursions (1) and (2) may be extremely complex – even in the absence of offtake. The herd growth function may have normal compensatory, overcompensatory, depensatory and critical depensatory properties for similar herd sizes at different periods. There is no reason to believe that normal compensatory growth (which leads asymptotically to convergence to equilibrium values for both herd size and range carrying capacity) will be encountered in reality. Indeed, it is more likely that growth will be overcompensatory (leading either to convergence via damped oscillations, or to non-convergent oscillation). But it is also perfectly possible for the growth function to be critically depensatory (leading to the collapse of the herd) where carrying capacity falls sharply over consecutive periods. In general, change in the size of the herd will vary directly with change in the level of grazing pressure given by the ratio x_t/k_t. In general, that is, overgrazing will lead to a decline in the size of the herd. However, it is important to add that since the natural rate of growth of both herd and carrying capacity is assumed to fluctuate, and since negative values for α_t and β_t are admissible, this will not necessarily be the case.

There are two broad senses in which the range may be overgrazed: economically and ecologically. Economic overgrazing is discussed in Section 5.3 below. Ecological overgazing can also be defined in two rather different ways. Ecological overgrazing may be said to be either 'fundamental' or 'current'. Fundamental ecological overgrazing may be said to exist whenever the stochastic equilibrium level of grazing pressure exceeds the level of grazing pressure corresponding to the maximum sustainable yield of the range: that is, when $\lim_{t \to \infty} x_t/k_t > x_m/k_m$. This corresponds to the meaning given to ecological overgrazing by, for example, Barrett (1989a). Current ecological overgrazing may be said to exist whenever the current level of grazing pressure exceeds the level of grazing pressure corresponding to the maximum sustainable yield of the range: that is, when $x_t/k_t > x_m/k_m$. Given the method of solution adopted here, we shall tend to focus on the more limited current ecological overgrazing (hereafter referred to as 'ecological overgrazing').

In the absence of offtake/restocking, the difference $x_{t+1} - x_t$ will be negative if $a_t x_t(1 - x_t/k_t) < 0$, which will occur either if $(1 - x_t/k_t) < 0$ and $\alpha_t > 0$; or if $(1 - x_t/k_t) > 0$ and $\alpha_t < 0$. The first alternative implies that the future size of the herd will decline where pressure is increasing on range that is currently being overgrazed. The second option implies that future herd sizes may fall (due to disease or drought, say) where the range is not currently being overgrazed. Moreover, since $x_t+1 - x_t > 0$ if $(1 - x_t/k_t) < 0$ and $\alpha_t < 0$, future herd sizes may rise where the range is being currently overgrazed in an ecological sense, providing that herd pressure on the range is falling. Similarly, the carrying capacity of the range in the uncontrolled case may decline for various reasons. $\beta_t k_t(1 - k_t/k_c) - \gamma_t(x_t/x_m)x_t < 0$ if either $\beta_t > 0$ and depletion is greater than regeneration, or if $\beta_t < 0$. The fact that herd size may exceed the carrying capacity of the range in any one year is not, therefore, a necessary condition for declining carrying capacity in the next year. Current ecological overgrazing may not imply fundamental ecological overgrazing.

5.3 THE MANAGEMENT PROBLEM

The optimal size of the herd relative to the carrying capacity of the range – the optimal level of grazing pressure – is the solution to a problem of decision-making under incomplete information. To facilitate construction of this problem we may assume that only the ecological parameters of the system are not known with certainty. Output prices are initially assumed to be determined exogenously, and to be constant over time. This is, in fact, a reasonable assumption to make with respect to some agricultural products – beef exports on long term contracts under the Lomé convention, for example. In general, however, the volatility of agricultural prices is a key factor in the sustainability in agriculture, particularly in the wake of price liberalization, and we shall return to this problem later. Incomplete information in the model accordingly refers to ignorance about the time trends of the physical system. The optimal policy is one that maximizes the expected welfare deriving from pastoral activity over an infinite horizon through choice of the level of offtake, and subject to the properties of the physical system. Such a policy may be said to be sustainable if it preserves the options available to future resource users by ensuring that the resource base is not depleted over time in the face of climatic (and other) shocks.

More formally, the problem is to:

$$\max_{\{u_t\}} E \left[\sum_{t=0}^{\infty} \rho^t W(x_t, k_t, u_t) \right] \tag{5a}$$

s.t

$$x_{t+1} - x_t = \alpha_t \, x_t \, (1 - x_t/k_t) - u_t \tag{5b}$$

$$k_{t+1} - k_t = \beta_t \, k_t \, (1 - k_t/k_c) - \gamma_t \, (x_t - u_t) \tag{5c}$$

$$x_0 \, (>0) = x(0) \tag{5d}$$

$$k_0 \, (>0) = k(0) \tag{5e}$$

$$u_t \leq x_t$$

$$x_t, k_t \geq 0 \tag{5f}$$

where E denotes expected value, and where $\rho = [1/(1+\delta)]$ denotes a discount factor, with δ being the rate of discount. There is no explicit sustainability constraint involved in this infinite horizon version of the problem.[4] However, the link between the optimal policy and the sustainability of the use of the resources involved will become clear.

The solution to this problem comprises a decision rule which fixes the optimal offtake policy for the given parameter values, and for the current values of the state variables. The rule is derived for the stochastic equilibrium of the system, and is applied sequentially to the actual values of the state variables, as these are observed, in a form of closed loop, or feedback, control. Derivation of the decision rule depends on the mean values of the stochastic parameters and accordingly abstracts from the uncertainty (process noise) generated by the random variation of the growth coefficients, α_t and β_t, and the rate of depletion, γ_t.

Without specializing the welfare function further at this stage, we may identify the necessary conditions for the control sequence $\{u_t\}$ to be optimal at the stochastic equilibrium of the system. Defining the current value Hamiltonian in terms of the expected values of $\underline{\alpha}$, $\underline{\beta}$ and $\underline{\gamma}$

$$H(x_t, u_t, \lambda_t) = W(x_t, k_t, u_t) + \rho\lambda_{t+1} \{\underline{\alpha}\, x_t[1 - x_t/k_t) + k_t] - u_t\} \qquad (6a)$$
$$+ \rho\zeta_{t+1} \{\underline{\beta}\, k_t(1 - k_t/k_c) - \underline{\gamma}(x_t - u_t)\}$$

the first order conditions require that

$$0 = H_{ut} = W_{ut} - \rho\lambda_{t+1} + \rho\zeta_{t+1}\ \underline{\gamma}\, x_t/x_m \qquad (6b)$$

$$\rho\lambda_{t+1} - \lambda_t = H_{xt} = -W_{xt} - \rho\lambda_{t+1}\ \underline{\alpha}(1 - 2x_t/k_t) + \rho\zeta_{t+1}\ \underline{\gamma} \qquad (6c)$$

$$\rho\zeta_{t+1} - \zeta_t = -H_{kt} = -W_{kt}\, \rho\lambda_{t+1}\ \underline{\alpha}\, x_t^2/k_t^2 - \rho\zeta_{t+1}\ \underline{\beta}(1 - 2k_t/k_c) \qquad (6d)$$

$$x_{t+1} - x_t = H_{\rho\lambda\,t+1} = \underline{\alpha}x_t(1 - x_t/k_t) - u_t \qquad (6e)$$

$$k_{t+1} - k_t = H_{\rho\zeta t+1} = \underline{\beta}k_t(1 - k_t/k_c) - \underline{\gamma}(x_t - u_t) \qquad (6f)$$

$$x_0 = x(0) \qquad (6g)$$

$$k_0 = k(0) \qquad (6h)$$

$$u_t \in U$$

The maximum condition (6b) requires that at the optimal level of offtake the marginal direct benefit of livestock sales, W_{ut}, should be equal to the discounted intertemporal 'shadow price' of offtake, $\rho\lambda_{t+1} + \rho\zeta_{t+1}\gamma$. The breakdown of this shadow price is quite intuitive. The first part of the expression, $\rho\lambda_{t+1}$, is just the cost of offtake in terms of the growth of the

---

[4] In a finite horizon problem an arbitrary terminal value of these variables may be assigned.

herd. It is the discounted value of future livestock units forgone by drawing down the herd in the current period.[5] The second part, $\rho\zeta_{t+1}\gamma$, is the gain of offtake in terms of future herd growth, attributable to the effects of lower herd densities on range regeneration. The full expression $\rho\lambda_{t+1} + \rho\zeta_{t+1}\gamma$ is therefore the discounted net user cost of offtake. The maximum condition accordingly requires that the marginal direct benefit of livestock sales should be equal to the discounted net user cost of offtake.

The adjoint equations, (6c) and (6d), describe the evolution of the shadow prices of the state variables as a function of offtake policy and the dynamics of the physical system. If a steady state solution (stochastic equilibrium) exists, such that $\lambda_t = \lambda_{t+1}$, and $\zeta_t = \zeta_{t+1}$, for $t = k, k + 1, \ldots$, (6b), (6c) and (6d) may be used to define a steady state 'rule' for determining the optimal level of grazing pressure. To simplify, let us first define the ratios:

$$\kappa_t \equiv k_t/k_c \tag{7a}$$

$$\psi_t \equiv x_t/x_t \tag{7b}$$

$$\omega_{xt} \equiv W_{xt}/W_{ut} \tag{7c}$$

$$\omega_{kt} \equiv W_{kt}/W_{ut} \tag{7d}$$

The first two have already been discussed: (7a) denotes the ratio of current carrying capacity relative to the maximum carrying capacity; and (7b) is the index of grazing pressure. (7c) and (7d) give the ratios of the marginal costs of, respectively, livestock and carrying capacity, to the marginal benefit of offtake. Once we specialize the welfare function these ratios will have the natural interpretation of the real-product cost of livestock maintenance and carrying capacity. Solving for the steady state values of λ and ζ from (6c) and (6d), inserting these into (6b), and using (7) yields the quadratic equation:

$$0 = \psi_t^{*2}\underline{\alpha}\gamma(1+\omega_{xt}) - \psi_t *2 \underline{\alpha}[\beta(1-2\kappa_t) - \omega_{kt}\gamma - \delta] + \omega_{kt}\gamma(1 - \underline{\alpha} + \delta)$$
$$+ [\beta(1-2\kappa_t) - \delta][\omega_{xt} + \underline{\alpha} - \delta] \tag{8}$$

Ψ_t^*, a positive root of (8), denotes the optimal level of grazing pressure.

It is now possible to be more precise about the concepts of overgrazing already discussed. Two measures of overgrazing can be identified: a measure of economic overgrazing, ξ_t, and a measure of ecological overgrazing, χ_t. These are defined as follows:

---

[5] Since the herd is assumed to be homogeneous for purposes of this chapter, offtake of any livestock units has the same effect on future herd growth. This abstracts from the specialized roles of livestock units by age and gender. Naturally, in the management of both domestic livestock and wildlife, the direct future costs of offtake in terms of herd growth can be minimized by selection of the units to be withdrawn from the range.

$$\xi_t = (\psi_t/\psi_t^*) - 1 \qquad (9a)$$

$$\chi_t = (\psi_t/\psi_m) - 1 \qquad (9b)$$

where Ψ_m denotes the maximum sustainable grazing pressure – the grazing pressure at the maximum sustainable yield. Overgrazing in an economic sense may be said to exist only if the actual level of grazing pressure exceeds the optimal level of grazing pressure ($\xi_t > 0$), whereas overgrazing in an ecological sense may be said to exist only if the actual level of grazing pressure exceeds the maximum sustainable grazing pressure ($\chi_t > 0$). Notice that the optimal level of grazing pressure is not necessarily the same as the maximum sustainable grazing pressure. Whether optimal grazing pressure is greater or less than the maximum sustainable grazing pressure depends on the economic parameters of the system – the relative prices facing resource users. If relative prices are such that it is optimal to 'mine' the range, the optimal grazing pressure will exceed the maximum sustainable grazing pressure – implying that there will be fundamental ecological overgrazing. On the other hand if relative prices are consistent with the sustainable use of the resource, the optimal grazing pressure will be less than or equal to the maximum sustainable grazing pressure. We shall return to this point later.

The construction of a decision rule for the level of offtake from (8) requires one further step. The optimal grazing pressure in any given period is determined by reference to the mean value of the economic and ecological parameters of the system, and to the current carrying capacity of the range, k_t. Given ψ_t^* the optimal herd size corresponding to kt is obtained directly as $\psi_t^* k_t$, and the expected steady state offtake corresponding to this herd size is simply $\alpha \psi_t^* k_t (1 - \psi^*)$. For a system away from the steady state this requires some modification.

In the general case it may be expressed as:

$$u_t^* = \underline{\alpha} \psi_t^* k_t (1 - \psi^*)(1 + \xi_t) \qquad (10)$$

implying that offtake will be higher (or lower) than the steady state level in proportion to the degree of economic overgrazing (or understocking) in the system. Substitution of the optimal grazing pressure into (8) yields optimal offtake.

Equations (8) and (10) provide the basis for a sequential feedback control policy in which the level of offtake in any given period is determined for the observed values of the state variables, and for the expected values of the stochastic ecological parameters. If the bioeconomic system is dynamically stable, then grazing pressure in a system subject to such a control policy will tend towards the optimal level – the stochastic equilibrium of the system. As we shall see, however, whether the bioeconomic system is stable depends on the economic environment within which the resource is used. The characteristics of the control sequence under a range of economic conditions are illustrated in the following simulation.

5.4 A SIMULATION

To illustrate the construction of an optimal policy, we need to specialize the welfare function further. It is typically assumed that welfare in the rural economy is a function of farm profit. This implies that it takes the following simple additive separable form:

$$W(u_t, x_t, k_t) = pu_t - cx_t - rk_t \tag{11}$$

in which p denotes the constant slaughter price of offtake, c denotes the constant cost of livestock maintenance, and r denotes the constant cost of carrying capacity. r may be thought of as a productivity-related charge for the use of grazing land, or grazing fee. Since the main concern of this chapter is with the implications of a given economic environment for the optimal management of rangeland under ecological uncertainty, maximization of (11) is acceptable as a first approximation of the social objective function. There would, however, be very serious reservations about its use in the case of the pastoral economy in semi-arid low income countries.

First, the additive separable form of the function implies a neutral attitude towards risk that is inappropriate in these conditions. Second, it ignores transfers. In many cases of interest profit may be negative for most prices compatible with the sustainable use of rangeland – as is the case in many pastoral economies in sub-Saharan Africa where transfers account for a major part of rural incomes. Third, the assumption that the relative prices are constant over time is obviously too strong. Fourth, and most important, the role of livestock in pastoral economies in the low income countries goes far beyond the production of beef, sheepmeat or goatmeat. The maintenance of livestock does involve costs (which in reality are an increasing function of the level of grazing pressure), but it also provides significant benefits to its owners in the form of draft power, animal products, the status it confers, the insurance it provides against adverse climatic conditions, and the fact that it is 'privileged currency' in bridewealth and other important social transactions. What c approximates in (11), therefore, is the net unit maintenance cost of livestock. Given (11) it follows that the ratios ω_{xt} and ω_{kt} are just the marginal real-product cost of livestock maintenance, and the marginal real-product cost of carrying capacity.

This exercise is not intended to capture the behaviour of any given pastoral economy, even though the parameter values are based on data for the livestock sector in Botswana,[6] a good example of a case where control takes place in the context of an ecological system driven by fluctuating levels of rainfall. In this exercise, the sequences $\{\alpha_t\}$, $\{\beta_t\}$ and $\{\gamma_t\}$ are random numbers with means:

$$\underline{\alpha} = 0.24$$

$$\beta_t = 0.03$$

$$\underline{\gamma} = 0.06$$

and coefficients of variation

$$\sigma_\alpha/\underline{\alpha} = 0.65593$$

$$\sigma_\beta/\underline{\beta} = 0.25092$$

$$\sigma_\gamma/\underline{\gamma} = 0.2075$$

The value for the relative variance of $\{\beta_t\}$ derives from rainfall data for the Maun district, while $\{\alpha_t\}$ and $\{\gamma_t\}$ have the same relative variance as the average rate of growth of livestock in that country. These sequences are recorded in Table 5.1.

The initial conditions in the example have been selected to correspond to the case where a herd of unsustainable size is introduced to a range regenerating at the maximum rate. This implies that there is ecological overgrazing (in the sense that the current level of grazing pressure exceeds the level of grazing pressure associated with the maximum sustainable yield of the range). Initial values of x_t and k_t are:

$$x_0 = 50$$

$$k_0 = 100$$

The maximum carrying capacity of the range, k_m, is assumed to be 200. Given these values, the level of grazing pressure corresponding to the maximum sustainable yield of the range is $\psi_m = 0.335$, and the degree of ecological overgrazing is $\chi_t = 0.4992$.

The implication of these initial conditions is that if the herd were not controlled (if there was no offtake) ecological overgrazing would lead to a 'Malthusian' collapse of the herd. This is illustrated in Figure 5.1, which shows the time paths for the state variables, x_t and k_t, in the absence of any control ($u_t = 0$ for all t). In this and all subsequent figures, time is registered on the horizontal axis, the values on that axis being equal to $t+1$.

---

[6] See Botswana, Central Statistics Office (1988), 1985/6 Household Income and Expenditure Survey. Also Botswana, Ministry of Agriculture (1987). Current Botswana series do not provide direct estimates of rates of rangeland depletion or regeneration, so the mean values of β and γ reflect traditional differences in the duration of periods of land use and fallow in the semi-arid zones of sub-Saharan Africa (Allan, 1965).

Table 5.1 Values for $\{\alpha_t\}$ $\{\beta_t\}$ and $\{\gamma_t\}$

time ($t+1$)	α_t	β_t	η_t
1	0.37806	0.01885	0.04889
2	0.58355	0.02990	0.04309
3	0.27315	0.02342	0.04293
4	−0.01060	0.03973	0.06782
5	0.12203	0.02916	0.06052
6	0.36677	0.03096	0.08122
7	0.24842	0.03780	0.06003
8	0.07811	0.03282	0.04219
9	0.24325	0.04946	0.06188
10	0.36018	0.01359	0.07539
11	0.17217	0.02911	0.04828
12	0.38300	0.02151	0.04501
13	0.20607	0.03641	0.05170
14	0.12335	0.02297	0.06526
15	0.33795	0.02666	0.06523
16	0.14020	0.02733	0.04943
17	0.18917	0.03488	0.07395
18	0.24764	0.02719	0.07185
19	0.02067	0.01972	0.05583
20	0.11453	0.02607	0.07373
21	0.36579	0.03719	0.06513
22	0.04776	0.03320	0.06025
23	0.11209	0.02710	0.05620
24	0.39499	0.04084	0.05740
25	0.58355	0.02990	0.04309
26	0.27315	0.02342	0.04293
27	−0.01060	0.03973	0.06782
28	0.12203	0.02916	0.06052
29	0.36677	0.03096	0.08122
30	0.36677	0.03096	0.08122

To address the economic problem, we need to add information on the non-stochastic economic parameters. These have also been constructed from Botswana data, although no claim is made for their accuracy. The slaughter price of offtake, p, is taken as the numeraire. The constant cost of maintaining cattle on the range in terms of the slaughter price is assumed to be 0.05, and the implicit cost of accessing the range in terms of p is assumed to be 0.03. The control policy in this case is constructed sequentially. Given the initial value of the state variables, x_0 and k_0, the optimal offtake, u_t^*, is calculated from (8) (9) and (10) on the basis of the expected values of the stochastic ecological parameters, α, β, and γ, and the known values of the economic parameters, p, c and r. Using ex post observations on the actual values, α_0, β_0 and γ_0, and the equations of motion, (1) and (2),

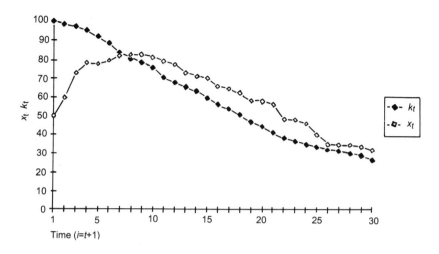

Figure 5.1 Time paths for x_t and k_t (uncontrolled case).

we may determine values for x_1 and k_1. These values are substituted into (8) to obtain ψ_1^*, and into (10) to obtain u_1^*, and so on.

Table 5.2 reports the values for the state and control variables over the first thirty periods in the control sequence, together with (undiscounted) values for the maximum and (farm profits), and the two measures of overgrazing in the system. The first point to note about this table is that at the parameter values assumed here the activity is sustainable over the time horizon of interest. The optimal grazing pressure converges to a value that is less than the maximum sustainable grazing pressure. The average value

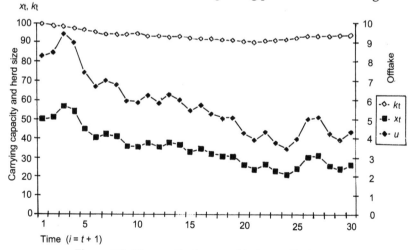

Figure 5.2 Time paths for x_t and k_t (controlled case).

Table 5.2 Time paths for state and control variables

Time	k_t	x_t	u_t	Y_t	x_t/k_t	ξ_t	χ_t
1	100.000	50.000	8.290	2.790	0.500	0.431	0.499
2	98.903	51.160	8.482	2.957	0.517	0.498	0.566
3	98.559	57.090	9.464	3.653	0.579	0.690	0.764
4	97.686	54.186	8.982	3.342	0.554	0.634	0.702
5	96.606	44.948	7.450	2.304	0.465	0.392	0.445
6	95.792	40.431	6.700	1.805	0.422	0.276	0.323
7	94.598	42.300	7.009	2.056	0.447	0.361	0.406
8	94.364	41.099	6.810	1.924	0.435	0.333	0.375
9	94.553	36.101	5.982	1.340	0.381	0.178	0.215
10	95.155	35.547	5.891	1.259	0.373	0.155	0.194
11	93.597	37.677	6.242	1.551	0.402	0.253	0.288
12	93.529	35.309	5.850	1.279	0.377	0.181	0.213
13	93.274	37.877	6.275	1.583	0.406	0.270	0.304
14	93.452	36.237	6.004	1.388	0.387	0.216	0.249
15	92.623	32.970	5.462	1.035	0.356	0.127	0.154
16	92.154	34.683	5.745	1.246	0.376	0.193	0.219
17	92.082	31.970	5.296	0.935	0.347	0.107	0.130
18	91.843	30.604	5.069	0.798	0.335	0.077	0.096
20	90.911	25.955	4.299	0.274	0.285	−0.072	−0.057
21	90.607	23.780	3.938	0.031	0.262	−0.140	−0.128
22	91.158	26.257	4.349	0.301	0.288	−0.060	−0.045
23	91.485	22.801	3.776	−0.107	0.249	−0.181	−0.168
24	91.761	20.942	3.469	−0.330	0.228	−0.246	−0.234
25	92.786	23.857	3.952	−0.023	0.257	−0.156	−0.139
26	93.416	30.247	5.011	0.696	0.323	0.053	0.076
27	93.499	30.822	5.106	0.760	0.329	0.071	0.094
28	93.733	25.497	4.224	0.137	0.272	−0.106	−0.087
29	93.897	23.537	3.900	−0.093	0.250	−0.172	−0.154
30	93.845	26.106	4.325	0.205	0.278	−0.082	−0.063

of χ_t is negative for $t > 19$. In other words, it is not optimal to 'mine' the range. Moreover, the net present value of the income stream is positive at a discount rate of 5% over the whole time horizon. Nevertheless, the activity is clearly marginal with farm profits at or near the stochastic equilibrium oscillating around zero.

The time paths for the state and control variables are shown in Figure 5.2. Carrying capacity and herd size are registered on the left hand vertical axis, and offtake is registered on the right hand vertical axis. The time paths for the state variables in this case are asymptotically convergent on their stochastic equilibrium values, the fluctuation in herd size being attributable to variation in the parameter α_t. Indeed, a control policy that is sensitive to change in current growth rates for the herd can significantly smooth the path of $\{x_t\}$.[7] Even with fluctuating herd sizes, however, this offtake policy effectively protects against degradation of the range.

5.5 THE ECONOMIC ENVIRONMENT AND ECOLOGICAL STRESS

The question that motivates the chapter is how the sustainability of resource utilization is related to the economic environment. More particularly, we are interested in the ecological impacts of a change in agricultural prices and income. This section first considers the relationship between the state variables and the system parameters of the general model, and then extends the simulation of section 5.4 to indicate the impact of specific change in the key economic parameters.

Recall that the sustainability of the ecological system underpinning pastoral activity implies its resilience to shock under the stress imposed by the level of that activity, where shock is the product of climatic perturbation and stress is indicated by the level of grazing pressure. In general, shocks may be due to either climatic or economic perturbation, but the latter case is not considered here.[8] The resilience of the ecological system is accordingly indicated by the response of the state variables (under an optimal policy) to perturbation of the ecological parameters. Non-resilience implies the collapse of the carrying capacity of rangeland, or the herd, or both.

The first point to make here is that the response of the system to exogenous shock does not depend on current levels of grazing pressure alone. From (8), (1) and (2), the change in the optimal level of grazing pressure in this period due to change in the ecological parameters in the last period is given by:

$$\frac{d\psi_t^*}{d\alpha_{t-1}} = -\frac{2\,\alpha[\psi_t^*\gamma(1+\omega_{xt}) - [\beta(1-2\kappa_\tau) - \omega_{\kappa\tau}\gamma - \delta]\,[x_{t-1}\,(1-\psi_{t-1})]}{k_t\,V}$$

$$\frac{d\psi_t^*}{d\beta_{t-1}} =$$

$$-\frac{[\psi_t^{*2}\alpha\gamma(1+\omega_{xt}) - 2\psi_t^*\alpha k_t\,(\beta(1-2\kappa_\tau) - \omega_{\kappa\tau} - \delta) + 2\beta(\alpha(2\psi_t^*-1) - \omega_{\kappa\tau} - \delta)\,(k_t^{\,2}/k_c)]\,[k_{t-1}\,(1-k_{t-1})]}{k_t^{\,2}\,V}$$

$$\frac{d\psi_t^*}{d\chi_{t-1}} =$$

$$-\frac{[\psi_t^{*2}\alpha\gamma(1+\omega_{xt}) - 2\psi_t^*\alpha k_t\,(\beta(1-2\kappa_\tau) - \omega_{\kappa\tau} - \delta) + 2\beta(\alpha((2\psi_t^*-1) - \omega_{\kappa\tau} - \delta)\,(k_t^{\,2}/k_c)]\,[\gamma_{t-1}\,(x_{t-1} - u_{t-1})]}{k_t^{\,2}\,V}$$

---

[7] Such a policy requires that $u_t^* = \alpha_t\,\psi_t^*\,k_t\,(1-\psi^*)(1+\xi_t)$, with α_t replacing α.
[8] Since the relation between economic and ecological parameters is perfectly symmetrical, nothing is lost by this restriction.

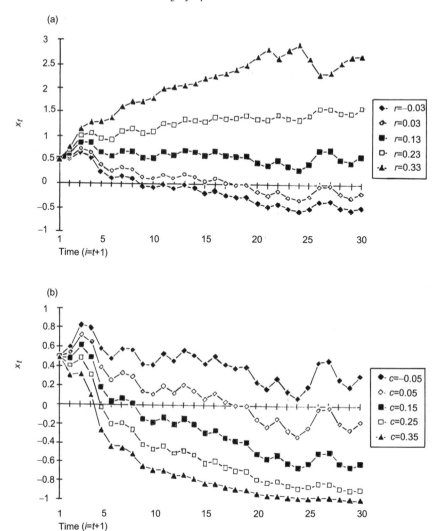

Figure 5.3 Time paths for $\{\chi_t\}$. (a) r variable, (b) c variable.

where

$$V = 2\underline{\alpha}[\psi_t^{*2}\,\underline{\gamma}(1+\omega_{xt}) - \psi_t^*\,k_t\,(\underline{\beta}(1-2\kappa_\tau) - \omega_{\kappa\tau} - \delta)] \qquad (12)$$

How the optimal level of grazing pressure changes as a result of perturbation of the ecological parameters at time $t-1$ depends, as one would expect, on the value of the state and control variables, k_{t-1}, x_{t-1} and u_{t-1} (and the mean value of the ecological parameters, $\underline{\alpha}$, $\underline{\beta}$, and $\underline{\gamma}$). But it also depends on the value of the economic parameters at time t, ω_{xt} and ω_{kt}: (the ratio of current input to output prices under the welfare function

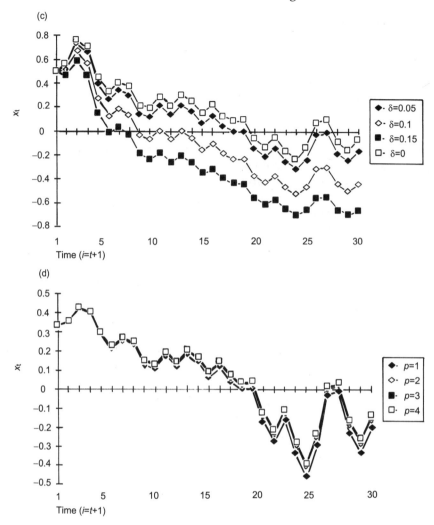

Figure 5.3 (continued) (c) δ variable, (d) *p* variable.

assumed here), and δ (the current discount or interest rate).

It is the economic parameters which determine whether the optimal level of stress on the ecological system is or is not sustainable. Since the economic parameters determine the optimal grazing pressure relative to the maximum sustainable grazing pressure, they also determine whether the stress imposed on the ecological system is sustainable in the absence of climatic shocks, and if it is sustainable in this sense, what magnitude of shocks can be absorbed without degrading the resource to the point where the herd collapses. Put another way, whether the optimal path for the state variables is stable in the face of climatic perturbation, depends on the value of the economic parameters relative to what may be termed the ecologically sustainable range. Given the dynamics of the ecological system, there exists a range for each non-ecological parameter over which the state variables are at least locally stable. This implies that change in relative prices or the rate of discount can either inhibit or enhance the resilience of the bio-economic system.

To illustrate these properties of the system, consider once again the simulation of section 5.4. The control policy maximizes welfare (farm profits) over an infinite time horizon, given a set of ecological and economic parameters. It has already been remarked that at the initial parameter values the economically optimal level of grazing pressure is stochastically sustainable, in the sense that it will not cause the collapse of the carrying capacity of the range. The application of the optimal control policy will not lead to positive ecological overgrazing. However, this is no longer true if the economic parameters are varied significantly from the initial values. Figures 5.3a–d describe the ecological overgrazing indicated by a range of economic parameter values, allowing both for positive and negative values. The inclusion of negative values covers the case where zero-priced pastoral inputs are subsidized – which is not as uncommon as might be supposed.

Discussion of the implications of this exercise is deferred to the next section. At this stage the results of the exercise are merely reported. Figure 5.3a shows the effects of varying the range rental or range user fee from the initial value of 0.03. The degree of ecological overgrazing is shown to vary directly with the range user fee: a range user subsidy resulting in the 'understocking' of the range, and an increase in range user fees raises the optimal level of grazing pressure. Indeed, in this example it turns out that a small rise in range user fees leads to a progressive fall in both the carrying capacity of the range and herd size under an optimal policy. Figure 5.3b shows, by contrast, that the effect of an increase in the maintenance cost of livestock is exactly the opposite: the degree of ecological overgrazing varying inversely with maintenance cost. Moreover, the lower the cost, the greater the relative variance in the degree of ecological overgrazing. In this case, the subsidizing of herd maintenance costs is associated with positive but highly fluctuating levels of ecological overgrazing. At the other extreme, very sharp increases in herd maintenance cost are associated with the progressive extinction of the herd through unsustainable levels of offtake. Figure 5.3c indicates a very similar relation between the discount (interest) rate and the optimal level of grazing pressure. Higher rates of discount encourage higher levels of offtake now, leaving herd densities lower in the future. Increasing rates of discount reduce the pressure on the range. Finally, Figure 5.3d indicates that the level of ecological overgrazing is not sensitive to change in the output price. Certainly, ecological overgrazing tends to increase as output prices rise, but in this example the increase is not significant over the range of prices tested.

The second measure of overgrazing of interest in this chapter is the level of economic overgrazing, ξ_t, which depends on the ratio of actual to optimal grazing pressure. The sequence $\{\xi_t\}$ describes the development of actual herd densities relative to the optimal density, and so sheds light on a rather different property of the bioeconomic system. Recall that $\{\chi_t\}$ describes the evolution of $\{\psi_t\}$ with respect to a fixed magnitude – the maximum sustainable grazing pressure. But the nature of the economic problem is such that the optimal level of grazing pressure is itself a function of the current state of the range and the herd. Consequently, $\{\psi_t{}^*\}$ evolves with the system, and may rise or fall as the state of the range or herd changes. The sequence $\{\xi_t\}$ accordingly indicates how the control process adapts to the evolution of optimal grazing pressure associated with the deterioration or enhancement of the carrying capacity of the range. But it also shows how differences in the optimal grazing pressure associated with differences in the economic environment impose greater or lesser stress on the natural environment, and so on the ability of the ecological system to accommodate climatic shocks.

Once again, these properties of the system are illustrated by extending the exercise reported in Section 5.4 – in this case to consider the effect on $\{\xi_t\}$ of a range of economic parameter values. The results are shown in Figures 5.4a–c. Figures 5.4a and 5.4b report the effect of changes in the value of range user fees and herd maintenance costs on the convergence of $\{\xi_t\}$, and illustrate the common feature of these experiments. Since different parameter values generate different optimal levels of grazing pressure, they will be associated with differences both in initial levels of economic overgrazing, and in the adjustment paths required to bring the ratio between the state variables into line with their optimal values. Because lower values of r are associated with lower optimal levels of grazing pressure, for example, they are also associated with greater initial economic overgrazing, and with greater required adjustments in the state variables. In addition, the more extreme the initial level of economic overgrazing (or understocking), the greater the change in the target values of the state variables along the way and the less stable is the convergence path for $\{\xi_t\}$. Figure 5.4c indicates the time paths for $\{\xi_t\}$ associated with a range of values for δ including one negative value. The positive values of δ reported here include only those for which there exist positive optimal levels of grazing pressure. Note that as δ rises, the optimal level of grazing pressure falls eventually becoming negative (passing through zero). Hence the initial level of economic overgrazing rises, eventually becoming negative (passing through infinity). For rates of discount higher than the maximum potential growth rate of the bioeconomic system, the herd is optimally extinguished over time. For the negative rate of discount reported here, the size of the herd

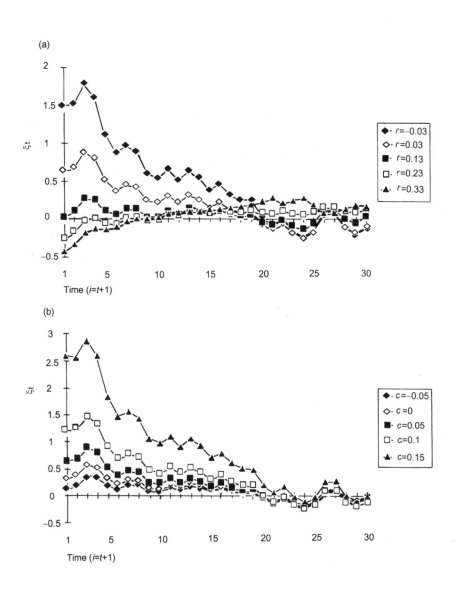

Figure 5.4 Time paths for $\{\varepsilon\}_t$. (a) r variable, (b) c variable

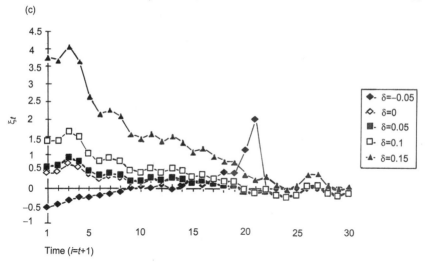

Figure 5.4 (continued) (c) δ variable.

relative to the current carrying capacity of the range is optimally increased to the point where the range collapses and the herd is similarly driven to extinction.

5.6 STRESS, SHOCK AND THE DISCOUNT RATE

To illustrate the implications of the stress levels imposed by different economic environments for the ability of the system to withstand shocks, it is convenient to focus on this last case – the discount rate. There is already considerable debate about the role of the discount rate in assuring the sustainability of resource use, and some concern about the ambiguous effects of a change in that rate. It is beyond the scope of this chapter to review these concerns, but they hinge on the balance between the myopia implicit in high rates of discount, and the demands on the stock of natural capital implicit in low discount rates. Pearce, Markandya and Barbier (1989), for example, are critical of myopia-based arguments against high rates of discount on these grounds. They argue that low discount rates may not lead to more sustainable use of renewable resources since they will tend to promote higher levels of investment (higher levels of economic activity) which may increase the stress on the environment. The model developed here supports the view that discount rates that are 'too low' relative to the productivity of the system may be incompatible with the sustainable use of resources, given that the rate of discount and the optimal grazing pressure are inversely related. But discount rates that are 'too high' relative to the productivity of the system will have the same effect. A progressive rise in the discount rate will ultimately lead to the collapse of the herd: a

The sustainability of optimal resource utilization

Table 5.3 Time paths for $\{\psi_t\}$: σ_α^2 variable

Time	$\sigma_\alpha^2 = 0$	$\sigma_\alpha^2 = 0.0225$	$\sigma_\alpha^2 = 0.09$	$\sigma_\alpha^2 = 0.2025$
1	0.500	0.500	0.500	0.500
2	0.482	0.506	0.520	0.383
3	0.463	0.432	0.467	0.284
4	0.448	0.458	0.630	0.268
5	0.436	0.569	0.646	0.203
6	0.425	0.460	0.497	0.287
7	0.419	0.485	0.411	0.211
8	0.408	0.505	0.435	0.189
9	0.398	0.445	0.475	0.305
10	0.387	0.419	0.420	0.429
11	0.387	0.416	0.429	0.489
12	0.380	0.372	0.348	0.416
13	0.374	0.358	0.390	0.264
14	0.367	0.299	0.432	0.298
15	0.365	0.332	0.419	0.340
16	0.362	0.341	0.361	0.433
17	0.358	0.302	0.229	0.484
18	0.355	0.311	0.250	0.348
19	0.354	0.381	0.349	0.331
20	0.352	0.321	0.265	0.431
21	0.351	0.334	0.222	0.454
22	0.347	0.298	0.156	0.622
23	0.344	0.323	0.170	0.711
24	0.341	0.359	0.144	0.568
25	0.337	0.375	0.162	0.330
26	0.333	0.350	0.126	0.239
27	0.331	0.354	0.142	0.338
28	0.329	0.326	0.191	0.464
29	0.327	0.313	0.198	0.111
30	0.328	0.216	0.238	0.167

progressive fall will ultimately lead to the collapse of the range.

Since the productivity of the system is a function of both the economic and the ecological environments, the stress imposed on the ecological system by a given discount rate will not depend on both the mean values of the stochastic ecological parameters, and on the values of the remaining economic parameters. However, whether or not a given level of stress is sustainable will depend on the variance of those parameters. In this example, the current values of the ecological parameters are assumed to be normally distributed about the mean, and the higher the variance of that distribution, the greater the probability of extreme events or 'climatic shocks'. The higher the variance of the stochastic ecological parameters, therefore, the less 'sustainable' is any given value of the economic parameters.

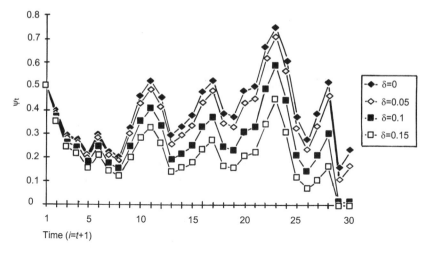

Figure 5.5 Time paths for $\{\psi\}$ at $\sigma_\alpha^2 = 0.2025$: δ variable.

Consider the effects of a change in the variance of herd growth rates on the tolerance of the system to different rates of discount. The first point to make is that higher variance in the ecological parameters implies higher variance in the optimal time path for the state variables under any economic environment. Table 5.3 reports values for $\{\Psi_t\}$ under different assumptions about the variance of α_t given the values of the economic parameters assumed in Section 5.4: that is, for $p = 1$, $c = 0.05$, $r = 0.03$, $\delta = 0.05$, $\alpha = 0.24$, $\beta = 0.03$ and $\gamma = 0.03$.

At these parameter values the fluctuations in grazing pressure associated with the higher variance in α_t are sustainable in the sense that neither of the state variables collapses. However, consider the effects of a change in the discount rate in an ecological environment characterized by the higher dispersion in the current values of α_t reported in Table 5.3. Figure 5.5 graphs the time paths for $\{\Psi_t\}$ under the same (non-negative) range of discount rates described in Figures 5.3c and 5.4c, but subject to a variance of $\sigma_\alpha^2 = 0.2025$ in the growth rate of the herd. As these two figures make clear, while higher discount rates are associated with optimal levels of grazing pressure well below the maximum sustainable level, the herd is not threatened with extinction under the original dispersion of $\{\alpha_t\}$. The same is not true under a sequence $\{\alpha_t\}$ with the higher variance.

In this case an optimal policy based on the expected rather than the actual value of α_t cannot contain fluctuations in the size of the herd. However, whether those fluctuations threaten the extinction of the herd depends on the optimal grazing pressure. At a discount rate of 10% the optimal grazing pressure is such that the herd is driven close to extinction within the time horizon of interest. At a discount rate of 15% the optimal grazing pressure is such that the herd is driven to extinction. Whereas the stress placed on the ecological system by an economic environment including a 15% discount rate is sustainable given the level of shocks expected under the original dispersion of α_t, it is unsustainable under the wider dispersion tested here. This suggests that considerable caution needs to be exercised in judging whether a given rate of discount (or a given value for any other economic parameter) is consistent with the sustainable use of environmental resources. It is the net effect of the economic environment that matters.

5.7 CONCLUSIONS

In recent years, considerable attention has been paid to the adverse environmental effects of the agricultural price regime in the low income countries. There is a very strong argument in the literature that the level of agricultural output prices is one reason for the degradation of arable and pastoral land in many low income countries, and that this is particularly true of sub-Saharan Africa. It is claimed that the agricultural sector in this region has been systematically disadvantaged by high levels of taxation of agricultural exports, financial repression, artificially depressed procurement prices and the protection of the industrial sector, and that this has prompted resource users to overexploit their natural environment. It is also claimed that liberalization of the agricultural price regime will contribute to the reversal of this trend. These claims raise a general problem that has not received much explicit attention. In what sense may an economic environment be incompatible with the sustainable use of natural resources drawn from sensitive ecological systems? Clearly, there are a number of relevant results in the existing theory of exhaustible and renewable resources. The 'iron law of the discount rate', for example, reflects a powerful result on the relationship between the rate of discount and the optimal rate of depletion of a resource. By implication, this provides a theory of the upper sustainable limit of an important economic parameter. There is no analogous implicit theory of the lower sustainable limit of the same parameter. Yet it is the suspicion that there exists a lower limit which has excited much recent concern in the literature. Nor is there any clear understanding of the way in which economic and ecological parameters interact to determine both the sustainable limits on parameter values, and the nature of the adjustment path given changes in parameter values.

This chapter considers how the implications of a given economic environment might be evaluated in the context of a bioeconomic model of resource use in a sensitive natural environment. At the most general level, the paper confirms that positive stochastic equilibrium values for the state and control variables may be identified only if the economic parameters of the system lie within ranges defined by the remaining economic and ecological parameter values. Parameter values outside of such ranges imply the collapse of the system. Moreover, the interdependence of the ecological and economic parameters means that change in one, changes the response of the system to the others. Put another way, at the most general level the chapter confirms that the economic environment sets the limits on the sustainable use of resources in any natural environment.

In the context of the model discussed here, it is possible to be more precise about the role of particular economic parameters. What is especially interesting, is that a number of parameters turn out to have unexpected effects. In so far as the improvement of procurement prices raises agricultural incomes, the argument that liberalization of the price system generates the means to conserve the resource base is well founded. However, the argument that higher slaughter prices in the pastoral economy will, *ceteris paribus*, encourage higher rates of offtake, and so lower levels of grazing pressure is more open to question. Higher slaughter prices have very little impact either on grazing pressure or on the degree of ecological overgrazing, but such impact as they do have is opposite to that expected in the literature. The higher the slaughter price, the greater the grazing pressure. The reason for this is that higher slaughter prices reduce both the real product cost of herd maintenance and the real product cost of carrying capacity, and these have opposite effects on grazing pressure. The implication of this is that unless the cost of carrying capacity (a grazing fee) is zero, higher slaughter prices will not lead to a reversal of the degradation of rangeland.

The different effects of a change in the cost of carrying capacity, r, and herd maintenance, c, may seem counterintuitive and so require some explanation. A fall in r implies that the marginal rate paid for range of a given carrying capacity falls relative to the cost of livestock maintenance on that range, inducing substitution in favour of range of higher carrying capacity. A fall in c, on the other hand, implies a fall in the marginal cost of livestock maintenance relative to the cost of carrying capacity, inducing substitution in favour of livestock. Hence a rise in grazing fees and a fall in livestock maintenance costs will each have the effect of increasing grazing pressure. This finding may be of particular interest in sub-Saharan Africa where r is at or close to zero in most cases, since it suggests that grazing fees may not be the way to secure a reduction in grazing pressure – unless they are inversely related to the carrying capacity of the range.

The impact of a change in discount rates is less unexpected, higher discount rates encouraging higher offtake, so leading to lower grazing pressure. As has already been remarked, this implies that excessive values of δ may lead to the extinction of the herd, while insufficient values of δ may lead to the collapse of the range through overstocking. As with the other economic parameters, the optimal solution to the economic problem will only be sustainable if δ falls within the admissible range. The point here is that the stochastic equilibrium of the pastoral economy is highly sensitive to its economic environment. It is assumed in this chapter that the ecological system is globally stable, implying that the carrying capacity of the range will always recover in the long run at a rate defined by β. But the economic system is not globally stable. Perturbation of any of the system parameters beyond the admissible range will cause the system to crash. Moreover, the admissible range will be greater or lesser depending on the variance of the stochastic parameters: the higher the variance in the stochastic parameters, the narrower the admissible range in the economic parameters.

REFERENCES

Allan, W. (1965), *The African Husbandman*, Oliver and Boyd, Edinburgh.

Barbier, E.B. (1988). 'Sustainable agriculture and the resource poor: policy issues and options', LEEC Paper 88–02, London Environmental Economics Centre, London.

Barbier, E.B. (1989a) The contribution of environmental and resource economics to an economics of sustainable development, *Development and Change*, **20**, 429–59.

Barbier, E.B. (1989b) Cash crops, food crops and sustainability: the case of Indonesia, *World Development*, **17** (6), 879–95.

Barrett, S. (1989a) 'On the overgrazing problem', LEEC Paper 89–07, London Environmental Economics Centre, London.

Barrett, S. (1989b) 'Optimal soil conservation and the reform of agricultural pricing policies', LEEC Paper 89–07, London Environmental Economics Centre, London.

Behnke, R.H. (1985) Measuring the benefits of subsistence versus commercial livestock production in Africa, *Agricultural Systems*, **16**, 109–35.

Bond, M.E. (1983) 'Agricultural responses to prices in sub-Saharan African countries', IMF Staff Papers, 30.

Botswana, Central Statistical Office. (1988) *Household Income and Expenditure Survey: 1985/6*, Gaborone, Government Printer.

Botswana, Ministry of Agriculture. (1987) *1986 Botswana Agricultural Statistics*, Gaborone, Government Printer.

Charney, I. (1975) Dynamics of deserts and drought in the Sahel, *Quarterly Journal of the Royal Meteorological Society*, **101**, 193–202.

Clark, C.W. (1976) *Mathematical Bioeconomics: The Optimal Management of Renewable Resources*, John Wiley, New York.

Cleaver, K. (1985) 'The impact of price and exchange rate policies on agriculture in sub-Saharan Africa', World Bank Staff Working Paper, 728.

Cleaver, K. (1988) 'The use of price policy to stimulate agricultural growth in sub-Saharan Africa', Paper presented to the 8th Agricultural Sector Symposium on Trade, Aid, and Policy Reform for Agriculture.

Conway, G.R. (1987) 'The properties of agroecosystems', *Agricultural Systems*, **24** (2), 95–117.

Conway, G.R. and Barbier, E.B. (1990) *After the Green Revolution: Sustainable Agriculture for Development*, Earthscan, London.

Dyson-Hudson, N. (1984) 'Adaptive resource use by African pastoralists', in *Ecological Practice*, (eds Di Castri *et al.*), UNESCO, Paris.

ILCA (1978) *Mathematical modelling of livestock production systems: application of the Texas A. & M. production model in Botswana*, ILCA, Addis Ababa.

Lutz, E. and El Serafy, S. (1989) Environmental and resource accounting: an overview, in *Environmental Accounting and Sustainable Development*, (eds E. Lutz and S. Serafy), World Bank, Washington, pp. 1–7.

May, R.M. (1974) *Stability and Complexity in Model Ecosystems*, Princeton University Press, Princeton.

Pearce, D.W., Barbier, E.B. and Markandya, A. (1988) 'Environmental economics and decision-making in sub-Saharan Africa', LEEC Paper 88–01, London Environmental Economics Centre, London.

Pearce, D.W., Markandya, A. and Barbier, E.B. (1989) *Blueprint for a Green Economy*, Earthscan, London.

Perrings C.A. (1989a) 'Debt and Resource Degradation in Low Income Countries: The Adjustment Problem and the Perverse Effects of Poverty in Sub-Saharan Africa', in *Economic Development and World Debt*, (eds H. Singer and S. Sharma), Macmillan, London, pp. 321–34.

Perrings C.A. (1989b) An Optimal Path to Extinction? Poverty and Resource Degradation in the Open Agrarian Economy, *Journal of Development Economics*, **30**, 1–24.

Perrings C.A. (1989c) 'Industrial growth, rural income, and the sustainability of agriculture in the dual economy,' LEEC Paper 89–11.

Rao, J.M. (1989) Agricultural supply response: a survey, *Agricultural Economics*, **3**, (1), 1–22.

Repetto, R. (1986) *World Enough and Time*, Yale University Press, New Haven.

Repetto, R. (1989) 'Economic incentives for sustainable production', in *Environmental Management and Economic Development* (eds G. Shramme and J.J. Warford), Johns Hopkins for the World Bank, Baltimore, pp. 69–86.

Turner, R.K. (ed.) (1988) *Sustainable Environmental Management*, Bellhaven Press, London.

Walker, B.H. and Noy-Meir, I. (1982) Aspects of the stability and resilience of savanna ecosystems, in *Ecology of Tropical Savannas*, (eds B.J. Huntley and R.H. Walker), Springer-Verlag, Berlin, pp. 577–90.

Warford, J.J. (1989) 'Environmental management and economic policy in developing countries', in *Environmental Management and Economic Development*, (eds G. Schramme and J.J. Warford), Johns Hopkins for the World Bank, Baltimore, pp. 7–22.

Westoby, M. (1979) 'Elements of a theory of vegetation dynamics in arid rangelands', *Journal of Botany* (Israel), **28**, 169–94.

6

Economic and ecological carrying capacity: applications to pastoral systems in Zimbabwe.[1]

Ian Scoones

6.1 INTRODUCTION

The term 'carrying capacity' (CC) is the source of much confusion. This chapter will hopefully clarify some of the issues surrounding the distinctions between economic and ecological carrying capacity, making the implications for the policy debate clearer.

The discussion is based on the findings of field work carried out in the communal areas (small scale farming areas) in Zvishavane District, Zimbabwe since 1986. The study area is in the dry part of the country with annual rainfall ranging from *c.* 300 mm to 700 mm. A mixed agropastoral system combines dryland cropping (maize, millets, sorghum, groundnuts, sunflower, etc.) with livestock production (cattle, goats, sheep, donkeys, chickens). This chapter concentrates on the role of cattle as important inputs to the agricultural production system.

6.1.1 Contradictory policy thinking

The inherent contradictions of thinking on CC are becoming increasingly apparent with the revival of grazing management schemes in the commu-

[1] This chapter is based on a seminar paper delivered at the University of Zimbabwe and an Overseas Development Institute Pastoral Development Network Paper (Scoones, 1989a).

Economics and Ecology: New frontiers and sustainable development.
Edited by Edward B. Barbier. Published in 1993 by Chapman & Hall, 2–6 Boundary Row, London SE1 8HN. ISBN 0 412 48180 4.

nal areas over the last few years. These are management schemes based on the assumptions of the beef ranching model, with controlled livestock stocking rates, fixed, fenced paddocks and rotational patterns of grazing. This policy is being vigorously pursued in the dryland CAs by government and donors in the name of 'modernization' of the traditional livestock sector and environmental protection of the resource base, which is assumed to be endangered (Scoones, 1990).

In Mazvihwa communal area (CA), Zvishavane District, Agritex (the state agricultural extension agency) has planned a grazing scheme in one of the wards. The early planning documents for a subsection of the ward give the following data (Agritex, 1986):

> Area of scheme 4560 ha
> Number of households 350
> Number of Livestock Units (LSU) 1753
> Acceptable stocking rate (CC) 1 LSU: 9 ha

What are the implications of this? The data shows that the current stocking rate is 1 LSU (Livestock Unit) : 2.6 ha with 5 LSU per household. When the data was collected, livestock populations were still recovering from the devastating effects of the early 1980s drought. In general in southern Zimbabwe, many households are recognized to be suffering severe drought shortages. This necessarily impairs agricultural production – the primary source of livelihood for CA residents.

However, the grazing scheme plan proposes an 'acceptable' stocking rate equivalent to 0.69 LSU per household. In other words, the choice of one cow or one donkey or three and a half goats for each home. This would clearly not be economically sustainable. Talk of reducing stocking rates to 'acceptable' levels is not new in Zimbabwe. Concern has been expressed about perceived overstocking for a long time (e.g. Watt, 1913). Native Commissioners and government technical officers continually referred to the disastrous consequences that were imminent if carrying capacity levels were not attained and regularly bemoaned the 'low productivity' of African stock. A few examples drawn from archival records of the study area give a picture of early perceptions:

> Many cattle in the Lundi Reserve are so starved as to be unsaleable ... this state may be attributable to lack of pasturage in this poor and overstocked reserve (Native Commissioner (NC), Belingwe [Mberengwa], Annual Report, 1925).

> Overstocking ... and a host of other causes have all accelerated the destruction of our natural resources. (Assistant Native Commissioner (ANC) Shabani [Zvishavane], 1945).

> (The grazing areas) have by continuous heavy grazing every growing

season, year after year been pushed to a point from which, if relief is not afforded very soon, deterioration will become increasingly rapid. (Dr O. West, Pasture Officer's report after a visit to Belingwe Reserve, 1948).

Early attempts at government intervention in livestock development in African areas were concentrated during the 1920s and 1930s on breed improvement. This gave way to the destocking policy of the following two decades where government tried to intervene directly and reduce stock to 'acceptable levels'. This was initially under the auspices of the Natural Resources Act (1942) and later under the Native Land Husbandry Act (1951). By the 1960s compulsory destocking had been abandoned largely due to political opposition, in favour of stock control in the context of rotational grazing schemes (Scoones, forthcoming).

The experience of attempts to control stocking rates to fixed levels is relevant here. Two issues make a recurrent appearance in reports and recollections of the time. Firstly, there were doubts about the validity of the CC assessments.

The NC for Gutu in his 1944 report comments:

So much has been said about overstocking in this district after the most cursory examination by folk deemed to have expert knowledge.

In a similar vein, the Land Development Officer for Gwanda comments in a letter to the Native Commissioner in 1948:

The development caused by hasty decisions made after hurried tours where little or nothing of the particular area is known might be disastrous ...

He goes on to warn against rushing into destocking, grazing management schemes, and a proposed programme of bush clearing. People recall the destocking in Mazvihwa. One old man remembers how:

There was plenty of grass for the cattle. They just came to brand our cattle and make us poor.

This memory is typical of many that both question the technical arguments behind the destocking policy and regarded it as an assault on their livelihoods. This latter impression is the second recurrent issue in discussion of destocking. The Native Commissioner of Belingwe comments in a letter to the Provincial Native Commissioner of Gwelo [Gweru] in 1947:

One wonders if we are not doing the native cattle industry untold harm by cutting down his holdings to an uneconomic figure.

This was an early realization of the impossibility of pursuing a supposedly environmentally sound policy that undermined the economic livelihoods of the people it was directed at.

Current policy is essentially similar to those being pursued since the 1940s. The First Five Year Development Plan (1986–1990) states:

> The most important aspect of livestock production which is occupying the mind of government is the accelerating and continuous deleterious effects of overstocking and overgrazing in the communal lands which are causing severe and potentially irreversible ecological degradation ... A comprehensive national programme that focuses on these issues will be implemented ... Such a programme will include stock control, better land management and destocking where necessary. (Zimbabwe, 1986, 27).

The Zimbabwe National Conservation Strategy (Zimbabwe, 1987, 25) re-emphasizes the policy line:

> In many cases this may require that animal numbers are restricted to a stocking rate that does not suppress the perennial grasses in well managed areas and allows the grass cover to recover in degraded areas.

If current policy is to be successful, then a number of contradictions experienced by earlier policy makers must be faced up to. The most fundamental question is: how can the economic sustainability of livestock production be assured while maintaining ecological sustainability? To answer this we need to ask what are the determinants of economic and ecological sustainability in CA livestock production? What do we mean by 'carrying capacity', and what are the appropriate ways of assessing it? What aspects of the CA production system and CA environment can be developed to enhance sustainability? These questions will be the central themes of the following discussions.

6.2 THE PRODUCTIVITY OF CA CATTLE: WHAT IS THE ECONOMIC CC?

A useful distinction is that between economic and ecological CC (Caughley, 1983). Ecological CC refers to the maximum number of animals the land can hold without being subject to density dependent mortality and permanent environmental degradation. Ecological CC is determined by environmental factors. Economic CC is the stocking rate that offers maximum economic returns and is determined by the economic objectives of the producers, i.e. by their definition of 'productivity'. The distinction can be illustrated graphically (Figure 6.1).

How productivity is measured is a critical issue as this will determine the assessment of economic CC. There are a number of ways that the productivity of livestock systems can be measured. These can be recorded in terms of production per individual animal or per unit area (Behnke, 1985).

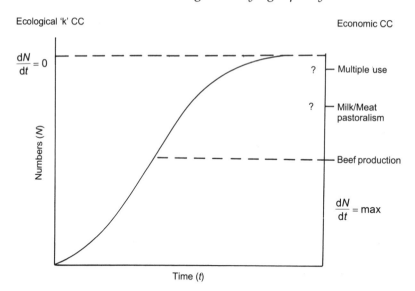

Figure 6.1 Economic and ecological carrying capacity (CC) under different production systems.

6.2.1 Biological measures for beef production

A beef producer's economic objective is to maximize the output of marketable meat. Productivity can be assessed in terms of a number of biological parameters that can be combined into a productivity index. Estimates are generally based on the weight of beef produced per cow per year calculated according to calving percentage, viability and pre/post weaning growth. There is a trade-off between productivity per animal and per unit land (Jones and Sandland, 1974) and the economic CC is at an intermediate stocking rate determined by the maximum economic returns per unit area.

6.2.2 Energetic or protein value measures

If the livestock enterprise is concerned with both the output of meat and milk, then a measure which combines the values of these products is required. Energy measures have been used to assess the productivity of pastoral systems in East Africa (Coughenour *et al*, 1985) and protein output has been used as a way of comparing the productivity of different livestock systems in Botswana (de Ridder and Wagenaar, 1986).

6.2.3 Economic measures

Where there are multiple economic uses of cattle then a monetary common currency can be used. In the case of subsistence production assigning

monetary value to outputs is problematic, but a broad indication of productivity can be arrived at by providing replacement costs (the value of
marketed substitutes) for those outputs not sold on the market (Behnke,
1985).

Assessments of CC used in Zimbabwe use beef production parameters
to estimate the stocking rate with maximum productivity (Kennan, 1969).
Rangeland indicators of this stocking rate are then used to assess CC (Ivy,
1969). This CC level is the economic CC for beef production which may
have little relevance to CA systems. The latter economic measure is most
applicable to CA cattle production, since the value of cattle is determined
by a range of outputs including use for draught, transport, manure and
milk production. They are also valued as capital assets and a relatively
stable investment.

A preliminary assessment of the productivity of CA cattle can be made
using the replacement cost method (Scoones, 1992b). This is based on an
analysis of data derived from a year long case study of 70 households in
Mazvihwa CA (Table 6.1).

The data highlights the high economic value of CA cattle. On the basis
of these, admittedly rather rough, calculations their 'productivity' appears
to be higher than that of commercial beef cattle that realize around $10/an-
imal/year in the same district. The comparatively higher returns realized
by CA cattle will be even larger if productivity/unit area is considered, as
stocking rates are considerably higher in the CAs.

The influence of the draught function on CA cattle production is such
that the economic CC will in turn be determined by a complex trade-off
between the economic advantages of more draught/work animals and the
effect of stocking rate on work ability, milk output, calf production and the
probability of death through poverty as stocking rate rises. The economic
CC will therefore tend towards the ecological CC limit (see Figure 6.1). It
will be advantageous to increase stocking rates to satisfy the demand for
draught, as shortages are currently due to the absolute lack of animals and
not so much because of distributional factors, which are evened out by
sharing and loaning practices.

High stocking rates in the CAs make economic sense. They are not the
result of irrational behaviour, poor management or backward attitudes. CA
livestock keepers are not beef producers; it simply does not make sense to
stock at economic CC levels designed for beef production. The question to
ask now is whether CA farmers' economic strategy is ecologically sustainable.

6.3 ECOLOGICAL SUSTAINABILITY: LIMITS TO LIVESTOCK NUMBERS IN CAs

How is CC assessed in CAs? It is instructive to look at recommended

Table 6.1 Economic value of cattle in the communal areas (Zimbabwe dollars)

Outputs	
Calving	Calving rate average 78%; calf death rate 25% in 80% of years and 65% in 20% of years. Value of calf $100 if sold.
Milk production	6 month lactation expect 480.6 litres at 50 c/l. Average economic output per cow $187.4/year
Work	At $30/day/span (41 spans, equal numbers of oxen and cows; average 2.8 cattle/span), total value of work per animal/year $623 (draught, $462 and transport $131)
Manure	Recoverable manure at 879 kg (2.6 carts)/adult/year (assumed less for years 1 and 2). With value at $10/cart, annual output is $26
Sales	Average sales price (1987) for sale at age 10 ($440 for oxen and $375 for cows)
Costs	
Livestock services	Veterinary services and dipping: total/animal/year at $7.3
Herding	Six months herding at $60/month; average herd of 8 animals gives $45/animal/year.
Discount rates	Discount rate of 10% assumed; plus adult mortality (3% in 80% of years and 10% in other years)

Discounted net present value of oxen and cows

Assumptions used in the calculations are shown above. The data are derived from field studies in Mazvihwa CA during 1986–7. Full details are contained in Scoones (1990).

1. Oxen

Net present value at 10% discount rate = $2280.
Internal rate of return = 137%

Year	Manure	Plough	Transport	Sale	Costs
0	6.5	0	0	0	152.3
1	13	0	0	0	52.3
2	26	462	131	0	52.3
3	26	462	131	0	52.3
4	26	462	131	0	52.3
5	26	462	131	0	52.3
6	26	462	131	0	52.3
7	26	462	131	0	52.3
8	26	462	131	0	52.3
9	26	462	131	0	52.3
10	26	462	131	440	52.3

2. Cow

Net present value at 10% discount rate = $3062.9
Internal rate of return = 173%

Year	Manure	Draught	Transport	Sale	Milk	Calf	Costs
0	6.5	0	0	0	0	0	52.3
1	13	0	0	0	0	0	52.3
2	26	462	131	0	187	53.8	52.3
3	26	462	131	0	187	53.8	52.3
4	26	462	131	0	187	53.8	52.3
5	26	462	131	0	187	53.8	52.3
6	26	462	131	0	187	53.8	52.3
7	26	462	131	0	187	53.8	52.3
8	26	462	131	0	0	0	52.3
9	26	462	131	0	0	0	52.3
10	26	462	131	440	0	0	52.3

Data derived from survey work in Mazvihwa CA, 1986–1987 (Scoones, 1990). The amounts are reported in Zimbabwe dollars (1987 exchange rate, US$ = c. Z$2).

procedures and the technical basis for estimating CC and ask whether these methods are appropriate to CA grazing lands.

The earliest assessments of CC carried out in Zimbabwe were based simply on assigning the area to one of 3 rainfall zones and the CC was given as 10 acres/beast for high rainfall, 13.33 acres/beast for medium rainfall and 16.66 acres/beast for low rainfall (Report of the Secretary for Native Affairs, 1947). This was established in a government notice of 1944 and emanated from the investigations of the Committee of Enquiry into the natural resources situation in 1942 (NRB, 1942), policy from South Africa, and early work at the Matopos Research Station (Pole-Evans, 1932; Hayle, 1932). The grazing area was calculated simply on the basis of the total area of the reserve less wasteland.

By the time of the grazing assessments for the Native Land Husbandry Act, the rules for assessing the area of available grazing land had become more elaborate, being calculated as 5/6 of the total usable land, which assumed that arable areas would be used for two months of the year. A further 7% would then be deducted to account for young stock not estimated in LSU calculations of stock numbers (Director of Native Agriculture reports 1951–53). The actual assessment of CC was still based on guesswork and experience of commercial ranching. Pasture assessment officers were sent out on tours when they made qualitative assessment of grazing condition and CC for different zones of each reserve. The visits were extremely cursory; for instance, Dr West visited Belingwe reserve (now Mberengwa CA) with a total area of 375 000 ha for a total of 6 days in the late dry season of 1948 to make his assessment. The pasture reports were then presented to the assessment committee for that area who would make the final decision on the stocking rate to be aimed for.

More recently attempts have been made to make grazing condition and CC assessments more rigorous. CC levels for different rangeland types have been indicated by the Matopos stocking rate trials (e.g. Kennan, 1969; Denny and Barnes, 1977; Dye and Spear, 1982) and in Intensive Conservation Area investigations. These have investigated the relationship between average weight gain per head and per acre at different stocking rates. The CC of a particular area is estimated by veld condition scoring (Ivy, 1969). Species composition, basal cover/density, vigour and forage production, litter and plant residues, and soil compaction and erosion are assessed and scores given out of a total of 50. The vegetation assessment is derived from the argument that a 'climax vegetation' is that which provides the best economic returns in beef production. 'Decreasers' are to be encouraged, while 'increasers' and 'invaders' are indicators of poor condition (Rattray, 1960). The condition percentage gives the assessor the recommended CC as a percentage of the potential as determined by stocking rate trials. For the driest areas (Region V), the rule of thumb appears to be 1 LSU : 10 ha, or 1:25 on degraded rangeland. The stocking level is then in practice determined by multiplying the assessed CC in LSU/unit area by the area of available grazing land (the practice of including an allowance for arable grazing is not part of current assessment procedures in Zvishavane at least). As an Agritex training resource puts its:

> It has been a question of trial and error by experienced people which have given us our guidelines, using sensible estimates based on the good farmers in a given region. From this figure, a somewhat arbitrary reduction is made from the potential grazing capacity to give a figure for current grazing capacity. (Anon/Agritex. Undated).

Vorster's comment made in 1960 is equally relevant today. He notes:

> There seems to be no definite information available on the extent to which the pastures may be stocked without causing erosion, and the stocking rates recommended for the reserves appear to be based on lower levels of veld utilization rather than on maximum carrying capacity. (Vorster, 1960).

Despite the increased complexity of assessment procedures, the underlying assumption is still that stocking rate trials measuring productivity with beef production parameters or experience of well run commercial ranches and the associated range indicators used are appropriate to CA situations. This can be questioned.

In beef ranching systems, economic objectives are different and the economic CC is at a low stocking rate, where a 'climax' herbaceous vegetation with low bush/tree cover may be optimal for the production system. This is not necessarily so in the CAs where higher stocking rates are economically desirable. A few examples will serve to illustrate this:

1. The replacement of perennials by annuals is regarded in conventional range management as a 'bad thing'. However, this may represent a shift in response to changing rainfall rather than an indicator of range trend (Dye and Spear, 1982). The presence of annuals may be advantageous in systems where protein deficiencies are a major constraint and where rapid responses to occasional rainfall sustains production (Penning de Vries and Djiteye, 1982).
2. 'Bush encroachment' is another indicator of poor range condition, but in most instances increased woody plants in dryland grazing areas are a definite advantage. Not only is browse crucial forage for all stock in the resource crunch periods of the dry season, but also some trees encourage valuable grass species such as *Panicum maximum* (Kennard and Walker, 1973).

What are needed are indicators that reflect ecological CC and do not translate the objectives of a particular production system into a picture of what the environment should look like in all situations. It is increasingly being realized that rangeland systems show a range of ecological dynamics – from equilibrium to non-equilibrium systems. This requires the development of appropriate management tools and rangeland indicators for each (Ellis and Swift, 1988; Behnke and Scoones, 1992). By confounding economic and ecological CC, existing assessment procedures may end up recommending stocking rates that undermine economic sustainability in CAs and do not directly address the key issue of ecological sustainability.

We need to know whether irreversible degradation caused by excessive stocking rates is occurring. With the possible exception of erosion assessments, the present veld condition measures are inadequate. It is not clear that the presence of 'sub climax' grassland or a high density woodland are indicators of permanent degradation.

A major obstacle in the way to finding out what ecological CC is for a particular area is that CC is not a fixed quantity and is not readily measurable. The very concept of ecological CC assumes an equilibrium system, yet in dryland areas where environmental variation creates large fluctuations in biomass production, vegetation composition and livestock population, non-equilibrium dynamics may apply. In such situations, livestock populations may never reach ecological CC levels as drought and other episodic events regularly knock back populations to lower levels. Under these conditions conventional range management indicators are largely inappropriate (Behnke and Scoones, 1992).

CC is thus a complex concept that is difficult to measure. Understanding the ecological dynamics of the system is vital to any assessment of ecological CC. Scoones (1993) shows that the populations of livestock in the Zimbabwe study areas fluctuate dramatically over time (see Figure 6.2). At some points over the period between 1923 and 1986 populations apparently reach an 'ecological CC' where birth rates equal death rates and populations remain static (assuming no movements in or out due to sales, slaughters, purchases). However regular droughts or other stress periods (e.g. cold spells) reduce populations to lower levels and for most of the period livestock populations remain below the ecological CC. The data therefore shows a mix of equilibrium and non-equilibrium dynamics over time, with density dependent regulation of population size operating intermittently with the dominant effect on population size being due to drought events.

6.4 INVESTIGATING LAND DEGRADATION

Soil erosion and vegetation change occurs with or without livestock and human interference. The important question is whether the environmental changes observed matter. The measurement of land degradation will thus depend on the economic objectives of the producer. A wildlife manager will be concerned with changes that reduce the biodiversity of the area; a

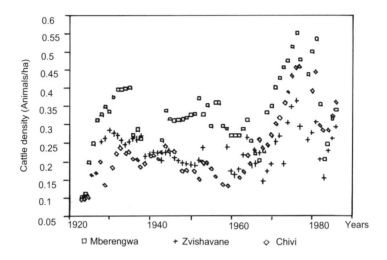

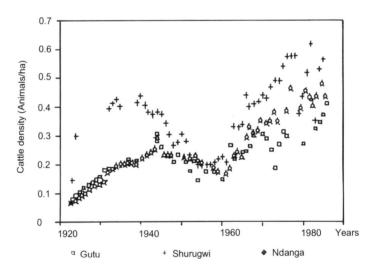

Figure 6.2 Populations of livestock in the Zimbabwe study areas fluctuates dramatically over time.

safari operator will be more concerned with the protection of large, valuable game; a beef rancher will be concerned with changes that affect the economic returns of beef production. In the CAs of Zimbabwe, changes that affect the economics of agropastoral production will be important. Abel and Blaikie (1989) provide a useful, economic definition of land degradation:

> Range degradation is an effectively permanent decline in the rate at which land yields livestock products under a given system of management. 'Effectively' means that natural processes will not rehabilitate the land within a timescale relevant to humans ... This definition excludes *reversible* vegetation changes even if these lead to temporary declines in secondary productivity.

Investigating degradation in CA livestock systems therefore requires the linking of environmental change with economic outputs. This can be done in a number of ways through the investigation of trends in primary or secondary production.

(i) Trends in primary production

Monitoring of primary production in semi-arid areas needs detailed study over very long periods since trends in potential productivity, (related to the long term ecological CC of land) are masked by the 'noise' created by widely fluctuating levels of actual productivity primarily determined by rainfall. No studies of this type have been done in the CAs of Zimbabwe. The universal perception of people from Mazvihwa is that no permanent decline is occurring (except in particular patches) and that as soon as rain falls again the grassland will be restored to previous levels just as in all previous drought cycles (Wilson, 1988; Scoones, 1992a).

(ii) Trends in secondary production

Short term monitoring of cattle production also fails to reveal any trends. Although Native Commissioners in their reports often commented on the poor state of grazing, deaths from poverty and the likelihood of imminent collapse (see earlier), these are interspersed by comments that contradict any claim that there is a terminal downward trend. A time series of qualitative observations for Mberengwa district illustrate the point.

> Considerable deaths occurred from either poverty or disease, particularly in the Lundi reserve that is overstocked. (Belingwe NC, 1930).

> On the whole grazing conditions have been shocking ... (Belingwe NC, 1934)

He [Cattle Inspector Gifford] is of the opinion that the reserve (Belingwe) is not overstocked and contained some of the finest native cattle he had yet seen. (Belingwe NC 1938)

The reserve [Lundi] is overstocked; as a consequence and aggravated by drought 1000 out of a total 1700 head died of poverty. (NC Belingwe, 1942)

Cattle continued despite the unprecedented drought in good condition [in Lundi reserve]. (NC Belingwe, 1947)

These observations refer to variations in actual productivity but because of its variability they can reveal little of trends in potential productivity.

A quantitative assessment can be attempted by looking at trends in population size and productivity parameters over time. Data for the period between 1923 and 1986 is derived from Native Commissioner returns and Veterinary Department records. This data (Scoones, 1990; Scoones, in press) shows that:

1. Between 1980 and 1986 there is no downward trend in total cattle population size. As this period shows no trend in rainfall levels, any trends could possibly be attributed to changing production potential of the land. The results indicate that the land is maintaining the population total.
2. Between 1960 and 1986 changes in birth and death rates are explained by changes in cattle density. Increased death rates and reduced death rates are shown with increased cattle population size over the period. There has been no long term increase in death rates and decrease in birth rates; changes are explicable in terms of total population density or rainfall (and so feed resource) variation.
3. Following the collapse in cattle populations during the 1982–84 drought, birth rates increased again to levels found during earlier periods. There has apparently been no irreversible change in cattle's capacity to reproduce.
4. Extraction rates (number of slaughtered and sold animals plus the change in herd inventory as a proportion of the total herd) are stable over the period from 1923 to 1986.

The results suggest that fluctuations in cattle populations and variations in production parameters are more due to variations in rainfall levels than changes in the intrinsic production potential of the land. In other words, just as Sandford (1982a) concluded, the available data on both primary and secondary production provide no firm evidence for irreversible land degradation. However this is not an excuse for complacency about patterns of environmental change in the CAs. It suggests that the study of land degradation requires a more sophisticated analysis. The following section attempts to disaggregate the issue and ask how cattle actually survive.

6.5 THE DETERMINANTS OF ECOLOGICAL CC: HOW DO CATTLE SURVIVE IN THE CAs?

In semi-arid systems animals survive by adaptive use of spatially and temporally heterogeneous resources. This is well documented for wildlife (e.g. McNaughton, 1985; Sinclair and Norton-Griffiths, 1979) and for no-madic pastoralists' herds (e.g. Dyson-Hudson, 1984). It is equally true for cattle in Zimbabwe's dryland CAs. A recognition of this brings us some way towards an understanding of what actually determines ecological CC. It also suggests appropriate directions for the development of livestock and grazing management in these areas. Two issues are particularly pertinent: the significance of drought induced movement of stock and the role of 'key resources' in sustaining livestock.

6.5.1 Drought movements of cattle

Movement in response to spatially variable forage production is a regular phenomenon in CAs. The movements observed in 1987 in Mazvihwa CA, as well as those reported for the 1982–84 drought from the Mototi Ward study area are shown in Figure 6.3.

The 1987 season resulted in localized shortages of forage in Mazvihwa. In April this was so for the cattle of Indava Ward; by August the cattle of Mototi Ward were beginning to be moved. During 1982–83, the lack of forage was more widespread. Cattle from Mototi Ward were moved into the hilly areas of Murowa and Mutambi Wards, before being moved on a large scale to areas of Chivi and especially to Mapanzure area in Runde CA via Zvishavane town. Some cattle moved as far as Shurugwi and Gweru to farms 150 km or more away from their home areas.

Movements have tended to be from areas of clay soil, drought suscepti-ble savanna (Mototi or Indava Wards) to areas of more stable production: the sandy soil zones of Murowa Ward, Chivi and Runde CAs. This macro-level patterning of resources is important in understanding the dynamics of cattle populations in southern Zimbabwe's CAs.

The loaning system (*kuronzera*) is central to the redistribution of grazing pressure in CAs. Cattle may be loaned on a temporary basis for the duration of the local crisis to relatives or friends or on a more long term arrangement which helps in reducing grazing pressure for the herd still resident at the owner's home. Large herd owners rarely keep more than 10–15 cattle in their home kraal but prefer to loan out to a number of '*miraga*' sites. This not only lowers the risk of local overgrazing but reduces herding and management requirements at the same time as assisting stockless relatives.

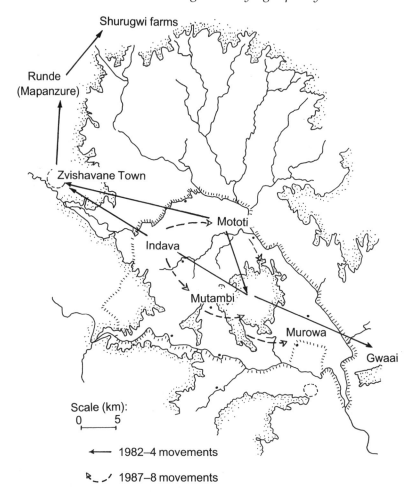

Figure 6.3 Movement in response to spatially variable forage production in 1987 and in the 1982–84 drought.

In the case of the 1982–83 crisis the situation was too extreme for local redistribution and the loaning system to cope. Many people adopted a transhumant existence for the duration of the movement, often living with their cattle in distant grazing areas or on commercial farms. Large scale movements of the type experienced in 1982–83 have been rare in the relatively recently settled area of Mototi Ward, Mazvihwa, although a similar exodus is remembered for the 1965 drought. In the nearby areas of central Chivi, people recall movements from the eutrophic plains areas into the hilly zones with their plentiful dambos (low-lying wetter patches) in the droughts of 1947, 1965, 1973 and 1982. In the arid CAs further south where grass production is even more variable and nearby refuges are

Table 6.2 Cattle survival and movement strategy (1982–4)
(i) Movement strategies employed

Strategy	Description
Strategy A	Movement out of Mototi Ward during 1982, usually around November
Strategy B	Movement during 1983 (mostly in the dry season between August and October)
Strategy C	No movement outside the area

(ii) Survival rates

Movement strategy	A	B	C
Survival%	40.1	22.9	3.3
Numbers	287	402	181

Statistics: $\chi^2 = 82.13$, df = 2, Significance = 0.0000

absent, NCs report having to make frequent arrangements for large scale migrations of cattle. For instance, the NC for Gwanda reports between 1925 and 1948 the necessity of moving thousands of cattle in 1938, 1941, 1942 and 1947.

Although many stock died in the 1982–84 drought, the toll was certainly reduced through the strategy of timely movement (see Table 6.2). Similarly, the local movements of 1987 offset the level of mortality.

6.5.2 Key resources

The use of the local grazing resource is equally patchy and adaptive. Much of CA cattle's feeding time, especially in the critical end of dry season period, is spent in small areas (perhaps 5% of the total grazing area); the rest of the grazing area is simply unused for most of the year. These small areas I shall call key resources. A key resource is a patch that offsets critical constraints either of forage quality (especially in sandy soil zones) or quantity (especially in clay soil zones). This has an important seasonal dimension that suggests that key resources – dambos, river banks and drainage lines – are critical in determining ecological CC.

A seasonal calendar for the utilization of components of the dryland grazing resource is presented in Table 6.3. This has been described by farmers from a number of areas and is supported by a detailed study of cattle foraging behaviour in Mazvihwa CA (Scoones, 1989b). It illustrates a patchy and temporally specific use of resources. The existing system is a form of rotational grazing influenced by the sequential availability of different resources, the reservation of the arable land grazing of the farming season and regulation through directed herding or foraging preferences.

Ecological CC is basically determined by the availability of fodder at the

Table 6.3 Seasonal calendar for grazing resource use by cattle in Mazvihwa CA

System type	Rainy season	Post-harvest	Mid-dry season	Late dry season
Clay soils eutrophic	G(KR)	A	A(KR)	KR/B
Sandy soil dystrophic	G(KR)	A	KR	B

Key: G = Extensive grazing area; A = Arable lands (stover and contour ridges); KR = Key resources (vleis, river banks, drainage lines and sinks); B = Browse.

close of the dry season particularly in rainfall deficit years: browse and key resources are the components that are critical. It is the availability of these resources and the facility of flexible utilization that has sustained high livestock populations in the CAs for a long period.

6.6 LIVESTOCK MANAGEMENT IN VARIABLE ENVIRONMENTS: SOME POLICY IMPLICATIONS

These observations have a number of important implications for policy. CA farmers prefer to follow an opportunistic stocking strategy: stocking at a high rate to ensure economic sustainability and by engaging in local destocking through movement when drought causes a collapse in available primary production. People prefer the tactic of movement to destocking through sales and later repurchasing animals because the low prices gained at the onset of drought do not, in their experience, allow repurchase at the end. In addition, a sales decision cannot be quickly reversed, whereas cattle can always be recalled to plough and graze locally if rains unexpectedly fall. Movements can therefore track the environment more effectively than adjustments through the marketing system.

Sandford (1982b) has compared two extreme types of stocking strategy. At one extreme is an opportunistic strategy that perfectly tracks environmental fluctuations adjusting stock levels to actual productivity. At the other is a conservative strategy that aims to keep stocking rates at a stable level in line with actual production in deficit years. In more variable environments, the costs of underutilization under the conservative strategy rise and an opportunistic strategy is increasingly favourable.

A conservative strategy is usually recommended: it may be 'safe', but the lower stocking rate it entails may undermine the economic sustainability of a CA system. On the other hand, a perfect opportunist strategy is infeasible; exact environmental tracking is not possible so some costs of temporary overstocking will be incurred. These trade-offs need to be closely examined. It seems that for CAs, drought planning should aim at encouraging opportunistic responses with contingencies being made for movements of cattle. Restrictions on such flexible and locally specific responses imposed by fenced grazing schemes or veterinary controls should be carefully examined in this context lest they reduce the sustainability of the system.

If the CC is to be increased key resource areas that should be focused on, as they directly determine CC levels. They are the most valuable components of the grazing system and should be where conservation efforts are first concentrated. Improvements in the extensive grazing land, although beneficial, will have a less direct effect on CC, and because it represents a far larger area, any attempts at intervention will be financially and operationally more difficult than a focused approach.

With draught provision being a primary objective for CA livestock development, key resources that provide end of dry season fodder are vital. More selective use of key resources opens up the possibilities of their use for selective feeding (e.g. draught animals/milking cows), reserved dry season grazing and their development as fodder banks.

6.6.1 Economic and ecological issues in the design of grazing schemes

Agropastoral systems are different in a number of important respects to commercial beef ranching. Each of these differences derives from the influence of alternative economic objectives on attitudes to stocking rates. Therefore, interventions applicable to commercial systems do not necessarily transfer to the CA situation. In low stocking rate commercial systems, forage consumption rarely exceeds actual primary production in non-key resource areas. Therefore, livestock can use the grazing resource in a relatively spatially uniform manner; key resources are not so critical. Paddocking systems that divide up the land with paddocks of comparable size are an effective way of managing the grazing land and the herd. This is in contrast to high stocking rate systems of CAs where actual production in non-key resource areas is rarely sufficient for dry season requirements.

A much more sensitive design of grazing management is required that allows flexible and adaptive use. Many of the grazing schemes currently being designed for the CAs do not fulfil these requirements. Paddock design rarely acknowledges the patchiness of available resources under high stocking rates. Applying a more or less uniform grid of paddocks to a highly heterogeneous environment, made more so by high levels of use, will mean that grazing schemes become quickly unsustainable. Farmers widely recognize this problem, but presently planners do not (Scoones and Cousins, 1991).

One farmer in Mazvihwa comments:

What is the point of putting the wire across the bare land; the cattle will just die with the coming of the paddocks.

Table 6.4 A comparison of grazing scheme design

	Multi-paddocked and fenced grazing schemes	Key resource grazing schemes
Stocking rates	Low	Relatively high
Economic sustainability	Increased beef production; reduced draught available; herding labour reduced	Maintain draught, manure, milk functions possibilities for selective feeding. Herding required
Ecological sustainability	Rotational resting improves grass in extensive grazing areas; movement constrained in droughts	Concentration on most degradation susceptible and valuable areas; free movement possible
Institutional Implications	Ward level planning by extension service; boundary disputes common	Local level planning possible, but untried
Cost	High (around Z$ 0.25 million per ward for fencing alone in 1987)	Lower; unknown

In order to make the paddock schemes function, destocking to commercial levels may be the only option as it is generally assumed that rotational grazing operates best at low stocking rates. This policy though, as has already been discussed, is not economically desirable.

How might appropriate grazing schemes be designed? One option is to focus on key resources. A possible strategy would be the identification of key resources for particular communities, fencing them off and establishing a system of regulated use. Any grazing management scheme will be locally specific, dependent on the critical constraints (e.g. quantity vs. quality of forage), the availability of forage in non-key resource areas, the browse resource and the objectives and management abilities of the local community. It is essential, therefore, that planning is locally based with farmers, with local knowledge of resources and their use, being the primary participants in design and development.

Table 6.4 compares 'key resource schemes' with conventional fenced grazing schemes. It can be seen that the principal characteristics of key resource schemes are that they are in line with CA farmer objectives (high stocking rates and the provision of draught animals) and are aimed directly at improving ecological CC.

6.7 CONCLUSIONS

This discussion has tried to show how the confusion over the term CC has arisen and attempted to unravel some of the contradictions in order to shed light on current policy dilemmas. In the past, the confusion has resulted in inappropriate measures of productivity and CC being applied to the CA situation. These have resulted in environmental policies which are not in line with local economic objectives. Such policies inevitably fail.

CAs will always be high stocking rate systems because of the multipurpose nature of cattle production. Having a high economic CC makes economic sense. In order to find development strategies that ensure economic and ecological sustainability in tandem, we need to look at how CA grazing systems are actually managed and concentrate on those factors that can maintain an economically viable stocking rate at ecologically sustainable levels. Two factors have been highlighted: macro level use of resources by adaptive movement, and temporally and spatially specific use of 'key resources'. These two factors ought to be the cornerstone of the design of grazing policies. Currently they are largely ignored in development attempts, since the focus is on the transfer of commercial ranch management systems to CAs. These are not appropriate as they assume different production objectives and are based on technical criteria that are open to question.

ACKNOWLEDGEMENTS

Research for this chapter was supported in part by an SERC/ESRC Studentship at Imperial College, London and the International Institute for Environment and Development. Field research was greatly assisted by K. Wilson, J. Madyakuseni, A. Mawere, B. Mukamuri, Z. Phiri, F. Shumba and many others from Zvishavane district.

ARCHIVAL SOURCES

Report of the NC Belingwe for the years ended December 1925, 1928, 1930, 1934, 1938, 1942, 1947. National Archives.
Report of the ANC Shabani for the year ended December 1945. National Archives.
Report of the NC Gutu for the year ended December 1944. National Archives.
Report of the Secretary for Native Affairs, Chief Native Commissioner and Director of Native Agriculture, 1947. National Archives.
Natural Resources Board: Native Enquiry (Oral evidence) 1942. National Archives S 988.
Director of Native Agriculture S1/53. Technical Survey party: method of procedure. January 1953. National Archives.
Pasture Officers Report, Belingwe Reserve. Dr O West, 1948 (held at Matopos Research Station).
Letter to NC Gwanda 26 May 1948 from LDO Gwanda (held at Matopos Research Station).

REFERENCES

Abel, N. and Blaikie, P. (1989) Land degradation, stocking rates and conservation policies for the communal rangelands of Botswana and Zimbabwe. *Land Degradation and Rehabilitation*, **1**, 1–27.

Agritex. (1986) Planning document for Indava ward grazing scheme. Unpublished mimeo. Agritex, Zvishavane.

Anon (undated) Veld condition and grazing capacity assessments. Veld Management Resource 19, Agritex Training Manual.

Behnke, R. (1985) Measuring the benefits of subsistence versus commercial livestock production in Africa. *Agricultural Systems*, **16**, 109–35.

Behnke, R. and Scoones, I. (1992) *Rethinking Range Ecology: Implications for Rangeland Management in Africa*, Environmental Working Paper, **53**, Word Bank, Washington.

Caughley, G. (1983) Working with ecological ideas. Guidelines for the management of large mammals in African conservation areas (ed. A Ferrar) *SANSP Report*, No 69.

Coughenour, M.B., Ellis, J.E., Swift, D.M. *et al.* (1985) Energy extraction and use in a nomadic pastoral ecosystem. *Science*, **230**, 619–25.

Denny, R.P. and Barnes, D.L. (1977) Trials of multi paddock systems on veld 3. A comparison of six grazing procedures at two stocking rates. *Rhod. J. Agric. Res.* **15**, 129–42.

Dye, P.J. and Spear, P.T. (1982) The effects of bush clearing and rainfall variability on grass yield and composition in SW Zimbabwe. *Zimb. J. Agric. Res.*, **20**, 103–18.

Dyson-Hudson, N. (1984) Adaptive resource use by African pastoralists. *Ecology in Practice: Part 1 Ecosystem Management*, (eds F. DiCastri, F.W.G. Baker, and M. Hadley), UNESCO, Paris.

Ellis, J. and Swift, D. (1988) Stability of African pastoral ecosystems: alternate paradigms and implications for development. *Journal of Range Management*, **41**, 450–9.

Hayle, D.G. (1932) Preliminary notes on intensity of grazing experiment. *Rhod. Agric. J.* **19**, 641–5.

Ivy, P. (1969) Veld condition assessments. Proc of Conex Veld Management Conferences in Bulawayo, pp 105–12; also, Agritex Veld Management Resource 17.

Jones, R.J. and Sandland, R.L. (1974) The relation between animal gain and stocking rate. Derivation of the relation from the results of grazing trials. *J. Agric. Sci.* (Camb.), **83**, 335–41.

Kennan, T (1969) Review of research into the cattle–grass relation. Proc Conex Veld Management Conference, Bulawayo.

Kennard, D.G. and Walker, B.H. (1973) Relationships between tree canopy cover and *Panicum maximum* in the vicinity of Fort Victoria. *Rhodesian J. Agricultural Research*, **11**, 145–53.

McNaughton, S. (1985) Ecology of a grazing ecosystem: the Serengeti. *Ecological Monographs*, **55**, 259–94.

Penning de Vries, F.W.T and Djiteye, M.A. (eds) (1982) *La productivité des paturages Sahelians: une étude des sols, des végétations, et de l'exploitation de cette resource naturelle*, Pudoc, Wageningen.

Pole-Evans, I.B. (1932) Pastures and their management. *Rhodesian Agriculture J.* **19**, 912–20.

Rattray, J.M. (1960) The habit, distribution, habitat, forage value and veld indicator value of the commoner Southern Rhodesian grasses. *Rhodesian Agriculture J*, **57**, 424.

de Ridder, N. and Wagenaar, K (1986). Energy and protein balances in traditional livestock systems and ranching in eastern Botswana. *Agricultural Systems,* **20,** 1–16.

Sandford, S (1982a). Livestock in the communal areas of Zimbabwe. Report prepared for the Ministry of Lands Resettlement and Rural Development. ODI, London.

Sandford, S (1982b). Pastoral strategies and desertification: opportunism and conservation in drylands in *Desertification and Development: Dryland Ecology in Social Perspective*, (eds B. Spooner and H. Mann), Academic Press, London, pp. 61–80.

Scoones, I. (1989a). Economic and ecological carrying capacity. Implications for livestock development in the dryland communal areas of Zimbabwe. ODI Pastoral Development Network Paper 27b, Overseas Development Institute, London.

Scoones, I. (1989b). Patch use by cattle in dryland Zimbabwe: farmer knowledge and ecological theory. ODI Pastoral Network Development Paper, 28b. ODI, London.

Scoones, I. (1990) Livestock populations and the household economy: a case study from southern Zimbabwe. Unpublished PhD thesis, University of London.

Scoones, I. (1993). Why are there so many animals? Cattle population dynamics in the communal areas of Zimbabwe. In: *Range Ecology at Disequilibrium: new models of natural variability and pastoral adaptation in African savannas* (eds R. Behnke, I. Scoones and C. Keven). Overseas Development Institute, London.

Scoones, I. (1992a). Land degradation and livestock production in Zimbabwe's communal areas. *Land Degradation and Rehabilitation,* **3,** 99–113.

Scoones, I. (1992b). The economic value of livestock in the communal areas of Zimbabwe. *Agricultural Systems,* **39,** 339–59.

Scoones, I. (forthcoming). Land and cattle in Zimbabwe: exploring different perspectives in the livestock policy debate, in *A Question of Perspective: Re-interpreting Environmental and Social Relations in Zimbabwe* (ed. M. Drinkwater) James Currey, London.

Scoones, I. and Cousins, B. (1991) Key resources for agriculture and grazing: the struggle for control over dambo resources in Zimbabwe, in *Wetlands in Drylands. The Agroecology of Savanna Systems in Africa*, (ed. I. Scoones.), Drylands Programme, IIED, London.

Sinclair, A. and Norton-Griffiths, M. (1979) *Serengeti: Dynamics of an Ecosystem,* University of Chicago Press, Chicago.

Vorster, L. (1960). The influence of prolonged seasonal defoliation on veld yields. *Proceedings of the Grasslands Society of Southern Africa,* **10,** 119–22.

Watt, M. (1913) The dangers and prevention of soil erosion. *Rhodesian Agriculture J.* **10,** 5.

Wilson, K. (1988) Indigenous conservation in Zimbabwe: soil erosion, land-use planning and rural life. Paper presented to the African Studies Association conference, Cambridge, September 1988.

Zimbabwe Government (1986) *First Five Year Development Plan*, 1986–1990, Government Printer, Harare.

Zimbabwe Government (1987) *The National Conservation Strategy: Zimbabwe's Road to Survival*, Natural Resources Board, Harare.

7

Tropical forests and biodiversity conservation: a new ecological imperative

Ian R. Swingland

7.1 WHAT IS BIODIVERSITY?

Biodiversity, or biological diversity, is the number, variety and variability of living organisms. It is used today both as a loose description to embrace the richness and variation of the living world and politically within the context of economics and development for the purposes of national and international conventions and agreements. It has three related components at different levels of organization: genes, species and ecosystems. Genetic diversity is the basis of biodiversity and, although the possible combinations of gene sequences will probably exceed the total number of atoms in the universe, only a small fraction (<1%) of genetic material in higher organisms is outwardly expressed in the form and function of the organism (Thomas, 1992). Species diversity is commonly considered the measure of biodiversity: 1.7 million species have been described although estimates for the total number of extant species range from 5–40 million (even 100 million), mainly consisting of insects and microorganisms. However such estimates, and the numbers of species already identified, depend on how they were described as distinct by taxonomists and systematists – ecologically, morphologically, physiologically, genetically, mathematically? The need for commonly agreed names which define a distinct species is vital to conservation but much confusion and dissent can arise (see also Rojas,

Economics and Ecology: New frontiers and sustainable development.
Edited by Edward B. Barbier. Published in 1993 by Chapman & Hall, 2–6 Boundary Row, London SE1 8HN. ISBN 0 412 48180 4.

1992). A tropical forest tree supporting hundreds of invertebrate and microorganism species makes a significant contribution to biodiversity in contrast to a temperate montane tree which has no dependent species. This, and the fact that tropical forests contain more threatened mammal species (for example) than any other habitat type (Thornback and Jenkins, 1982), explains why there is an international focus on tropical forests rather than temperate habitats. And why the concept of 'megadiversity' (Mittermeier and Werner, 1990), species or taxa endemicity (Myers, 1990; Vane-Wright, Humphries and Williams 1991) built around scores of species richness of a country or region, is being used for prioritizing international funding support; even though the usefulness of species inventories is too limited for conservation management on site. With the vogue in the last decade for NGOs to concentrate funding on habitats (=ecosystems) rather than species, the concept of ecosystem diversity has arisen even though NGOs raise most of their funding using charismatic species. There is no satisfactory method of measuring ecosystem diversity although the relative abundance and type of species present would be relevant.

The need for an unequivocal and precise meaning of biodiversity, which is scientifically sensible, is imperative if we are to develop policy and programmes for the future, and make decisions about the present.

7.2 WHAT DO WE KNOW ABOUT BIODIVERSITY?

The definition of biodiversity affects objectives in management. One could easily promote a timber extraction, or non-timber forest product programme, which conserves species richness (i.e. numbers of species) at the expense of genetic diversity. Indeed a current research programme to stimulate or increase the range of harvested tropical tree species not currently in trade, to take the pressure off the currently over-exploited species, may be misguided. It may lead to increased genetic as well as species impoverishment when foresters expand the species they take, and select only the best and most mature specimens thus removing the most productive and healthiest genetic stock.

Species diversity increases as the habitat becomes warmer, wetter and lower in altitude; the zenith being tropical moist forests covering 7% of the earth with 50–90% of the known species. However it all depends how you measure diversity. Certain tropical marine reefs in Indonesia and plants in South Africa and Madagascar are at least as diverse if we look at the larger organisms only. But these places of high biodiversity are in a belt around the world between the tropics (of Cancer and Capricorn). The question is why are there these differences in biodiversity between the tropics and temperate areas? Are they because of human activity, gene loss, global climate and weather change – or something more fundamental concerning extreme and unstable environments which are more prevalent the nearer you get to the poles and the difficulty of speciation in these more hostile habitats? For example, global climate change predicts an increase in the total potential area of tropical forest but this will be constrained by human use (Leemans and Halpin, 1992).

The relationship between loss of habitat and loss of biodiversity has never been satisfactorily characterized and there is no predictive model for policy-making and planning. Clearly introduced species (which out-compete, or hybridize with, the endemic species) and disease, habitat fragmentation, shifting cultivation, development, logging, selective removal of commercial non-timber species, transmigration, hunting, roads and railways, pollution, population pressure, fuel, recreation, tourism, international trade in species and other human influences have a deleterious effect. Larger species will be affected more severely than small species as areas of forest are reduced. But we do not know enough about the changes in variety, number, and distribution of species as habitats are reduced or adversely affected by humans. Gross estimates are all we have (Table 7.1), none of which are particularly edifying.

Although the persistence of species depends on their being constituted as a metapopulation (= subpopulations) distributed among many patches (May and Southwood, 1990), and the abundance of a species found in modified habitats surrounding original fragments (i.e. in a metapopulation) is an excellent predictor of vulnerability to extinction (Laurence, 1991), metapopulation theory has limited application in determining priorities in conservation. Obviously knowledge of local patch (subpopulation) dynamics does not determine the dynamics of a species population, for instance, but high variance in the local density does tend to buffer the species population from extremes of local competition and thus the species population is more stable than the subpopulation dynamics (Rosewell, 1990). Also it is clear that persistence time for an endangered species is maximized using one large forest reserve rather than two smaller reserves which have the same total area; the evidence coming from models, but also from the fact that the present-day distributions of vertebrates isolated from conspecifics during the Pleistocene demonstrates that a single larger reserve would have been the more effective strategy (Wright, 1990). Metapopulation theory, together with existing models of minimum viable

Table 7.1 Estimated rates of species extinction based on forest area losses

%Global loss per decade	Period	Estimation method	Authority
4	1975–2000	extrapolation	Myers, 1979
2–5	by 2020	species–area projection	Reid and Miller, 1989
1–5	1990–2015	species–area	Reid, 1992

population size (previously applied to single species) and estimates of the minimum area required, should enhance the accuracy and usefulness of the predictions (Jenkins, 1992).

7.3 THE ROLE OF TROPICAL FORESTS AS 'SOURCES' OF GLOBAL BIODIVERSITY

The significance of tropical moist forests to the maintenance and use of biodiversity is limitless. As a resource and environmental stabilizer, a tropical moist forest is the most important ecosystem. It provides timber, non-timber products, tourism income, plant genetic material, raw materials for biotechnology, medicines, pharmaceuticals, food and shelter for its people. It prevents erosion, maintains soil fertility, regulates water run-off, moderates climate, reduces stream and river sedimentation which kills fish, stops nutrient wash-out which pollutes waterways, and fixes carbon dioxide. Its very importance as a 'goldmine' of biodiversity may yet be the cause of its destruction.

The trees themselves are the most important issue and the way they are used by humans. Tropical forest timber is a very valuable commodity, creates enormous foreign exchange, and is a major contributor to most developing countries GNP. But few if any forests have been managed for the sustainable production of timber (Poore *et al.*, 1989). Management systems range from forest block demarcation to be held in reserve until economically worthwhile to cut, to selective logging of suitable individual trees and replanting with commercially-desirable endemics to clear felling. The most profitable option is the last. More recently many environmental economists and conservation biologists have been promoting the non-timber product utilization of tropical forests which not only is sustainable, low technology but also involves local people in co-operative arrangements which provide a stable long term future for both the people and the ecosystem (e.g. tagua or vegetable ivory, Mittermeier, pers. comm.).

Increasingly tourism, centred on foreigners' fascination with tropical forests, has become feasible and commercially very attractive to local and foreign businessmen. The profit produced within the local economy, which is only a minor proportion of the total profit, rarely filters down to the local forest people or benefits the maintenance of the very environment the tourists come to see. Increasingly governments are trying to recover the situation by charging entry fees, encouraging arts and crafts in local communities, and gaining tighter control on on-ground operators. In some cases local people co-operate and both manage and control the resource themselves (e.g. Masai in the Ngorongoro/Serengeti National Park) but there are no examples in tropical forests since the value of the crop is too valuable for the governments to permit local control and is politically unacceptable (protected area products in Madagascar, Durbin, in press).

The development of monoclonal or high-yield cultivars of higher plants for intensive agriculture is leading to a dangerous reliance on genetically uniform crops. This form of agriculture is leading to more habitat destruction of tropical forests, the very source of the genetic variability which will be needed when viral or bacterial epidemics, or pest outbreaks (e.g. the DDT, frog, insect story), devastate these vulnerable modern genetic monocultures. Many wild forebears of these crop plants are struggling to survive, and numerous agencies are trying to hold stocks of viable living plants or seed.

The question of biotechnology and its inclusion as an article in the *Convention on Biodiversity* presented at the Earth Summit (UNCED) in June 1992 is of such significance that it effectively dissuaded the United States from signing the Convention since it gave rights and dues to the country of origin of the organism that was subsequently developed and made commercially successful elsewhere. The US President at the time refused to sign the Convention on the basis it would cause job losses in the USA. Forest microorganisms, so tractable to biotechnological development, play a major role in the evolution and diversification of such as nutrient supply through the breakdown of cellulose and lignins, and mycorrhiza around the roots of 85% of all vascular plants are crucial to the absorption of growth-limiting nutrients. Food networks of all life are dependent on microorganisms which also play roles in the maintenance of soil structures, biodegradation, biocontrol such as plant pathogenic microorganisms which can limit plants and entomogenous microorganisms which can limit insect populations (Hawksworth and Colwell, 1992). Microbial biodiversity in tropical forests will in the future be used by biotechnologists to improve nitrogen-fixing in legumes, enhance nitrogen-fixing using cyanobacteria (e.g. in *Azolla* used in rice field systems), biocontrol of insects, plants, diseases and weeds, and finally mycorrhizal fungi for improved plant growth and stress tolerance. The medicines from tropical forest species are common, and require no biotechnological modification, but the question of who benefits – the foreign developer, the government, the local businessman or the people who depend on this dwindling habitat, and who probably gave away a lifetime's experience and wisdom for nothing by describing the medicinal use of each species – is still rife. The famous instance of the rosy periwinkle from Madagascar which has a dramatic effect on childhood leukaemia and is now grown in bulk in the USA has never benefited Madagascar or the local indigenous people.

The tropical forests support unique cultural systems, societies of gatherers, which are not only an integral part of the forests, in balance with (and vital for) the stablility of the ecosystem, but also its most intelligent guardians, capable of being the most important asset for future sustainable and profitable use of tropical forests.

Tropical forest ecosystems have very few recorded extinctions. Most of the species lost occurred on islands (75% of world total) and as a conse-quence existed effectively as a single population, without metapopulations, susceptible to extinction (Jenkins, 1992). Island species' ecology makes them innately vulnerable to predators. The low recorded extinction rates for tropical forests (2–3% per decade) are surprising because habitat loss is faster than in any other ecosystems (Ehrlich and Wilson, 1991). However there are lag-time effects not only between habitat loss and actual extinction (particularly of long-lived species), but also in the detection and reporting of such events. Furthermore since perhaps only 4% of species within tropical forests have been identified it is possible that extinction rates are much higher than currently recorded.

7.4 INTERPRETATIONS OF BIODIVERSITY FOR PLANNING

The use of biodiversity to define priorities is becoming urgent not just for conservation investment purposes but because the very process itself (of specifying an agreed and usable technique for measuring it) carries with it 'glittering prizes'. Numerous NGOs, institutes, international agencies, and universities are trying to define, and have accepted, their method of quan-tifying biodiversity. So what are these methods and how are resources deployed in conserving biodiversity?

What measurement could be applied to decide which area, habitat or ecosystem should be included, as a priority, in any conservation pro-gramme?

1. It has the highest level of species' endemicity (Myers, 1990) or taxa endemicity, called critical faunas analysis (Ackery and Vane-Wright, 1984).
2. It has the highest number of species, i.e. species richness.
3. It is the most significant ecosystem of its type, decided on general perceptions or criteria, and is a necessary part of a national or inter-national 'portfolio' of habitats.
4. It contains and supports flagship species on which attention and funds can be focused.
5. It contains and supports *keystone species* on which the ecosystem itself depends, and the ecological integrity of surrounding areas.
6. It contains a species, deemed important by some other criteria (e.g. charismatic mammal which would attract funds, commercially im-portant), where all indications point to extinction using viability modelling (population viability analysis) within a relatively short (and specified) time without conservation measures being taken.
7. Population analysis using comprehensive raw data, and not esti-mates, indicates that the species is likely to enter a non-recoverable trend without management.

8.   It is highly tractable for long term conservation because the local people can easily be integrated and involved in management (and benefit from such involvement); local integration.
9.   The presence of species which can be sustainably used e.g. cropping, farming or ranching species, tourism.
10.   Political exigency.

Although the methods of conservation within the policy of a country may be driven by more pressing needs – family planning, education, politics, internal conflict, financial planning and investment, individual vested interests – the ten-point list above are ways in which current policy and decisions are being made. Each of the methods has drawbacks and while the list is not comprehensive some combination of the various approaches will be the most effective strategy.

Endemicity and species richness are useful starting points in defining priorities on the global level but without information on the possibility of extinction using viability modelling or population analysis the urgency of the position cannot be assessed. Moreover with the increasing emphasis on the integration of local people into conservation programmes to minimize longterm costs, and provide a more stable basis for the people and their natural environment, the potential for local integration, coupled with sustainable use, cannot be ignored. Since national or external funding will generally be involved in pump-priming most projects the importance of an area relative to others using an ecosystem diversity (or portfolio) approach will be needed. The presence of a flagship or keystone species will be significant in raising such funds. Clearly political exigencies or pure chance can enter the situation, and we have yet to articulate whether we should not be using genetic diversity as the key measure, but in the absence of realistic methods of quantifying these characteristics they must remain imponderable objectives for the moment.

The differing approaches being advocated for biodiversity conservation are not just guided by the available methodologies but are symptomatic of the underlying philosophies. The evolution-based approach is predominantly the preserve of biologists and is concerned with the maintenance of diversity as an unqualified objective unaffected by economics; whereas the need for conservation, and the uses of biodiversity – the resource-based argument – are what are used to 'sell' the proposition to decision- and policy-makers. Where they come together is in the ideal of sustainability and the methods of achieving it.

7.5 ISSUES CONCERNING BIODIVERSITY AND TROPICAL FORESTS

The predominant issue involving biodiversity and tropical forests centres on the right of use of this common resource. Who has the right to exploit the forests and in which way? The local people, the government, the businessmen, the international community (however defined) or the scientists? Can we effect long term sustainability since most of those making the decisions want short term profit?

The uses of tropical forests are enormous:

1. plant food crops (200–3000 spp.);
2. genetic resources for plant crop improvement (wild genes resistant to viruses and invertebrates, promote rapid growth, higher yields);
3. timber (world export 1989 US$ 6000m);
4. rattans (90% extracted from the wild, industry employs 0.5m people, >90% spp. vulnerable or endangered);
5. medicinal plants and animals (90 plant spp. in worldwide use, US annually import US$20m, 3217 animal spp., annual world trade US$350–500m);
6. ornamental use of plants and animals (species orchids, cacti, insects, skins, feathers, worldwide trade US$3000m);
7. meat and eggs (no estimates available, mostly locally consumed);
8. working animals (16 000 working Indian elephants in captivity);
9. sport hunting (annual worldwide turnover in tropical forests >US$43m);
10. tourism, recreation (annual worldwide turnover US$1300m);
11. captive animals as pets or for display (annual worldwide turnover US$1600–2300m);
12. animal domestication (crib-biting wear found on horses teeth 27 – 29 000BP (Blagg, pers. comm.), 3237 extant breeds (Mason, 1988) <211 originated from tropical forests, developing worldwide databank (Maijala, 1990));
13. selective breeding and livestock farming (quicker to import genes from the wild than selection within a breed, few tropical forest examples – kouprey *Bos sauvelii* when found, some pigs and peccaries).

The prizes are considerable but little of this wealth is re-invested in the tropical forests and their indigenes. In consequence it is being used as a non-renewable resource in a desperate scramble for instant wealth. To ameliorate, improve, and to provide attractive alternative means of management which are sustainable, one or two institutions are now transferring the know-how and appropriate simple technology to resolve the technical difficulties.

But the overwhelming issue arising out of the uses and values of biodiversity in tropical forests, the politico-economic market, will never be resolved. However there is a specific suggestion that might be an improvement. It is clear that for decades the traditional, confrontational method of dealing with conflicts in interests over natural resources does not resolve the problems. Most methods rely on one party or another fulfilling undertakings which can easily be broken without any framework of checks and balances which ensure consistency and homoeostasis.

An international consultative meeting of twenty people working on the environment was held at the United Nations Environment Programme (UNEP) headquarters in Nairobi 27–28 February 1992 to discuss the next decade, and the UN response to the debate on the environment and economics. They advised that problems should increasingly be solved by adopting international or regional conventions or protocols that enable different 'stakeholders' to come together and negotiate by trading rights to common property resources.

The exact nature of the convention or protocol should depend on specifics of the problem. Generally, however, they should entail a minimum standard that all can agree upon, and one which can be modified based on the accumulation of additional scientific data that will allow both effective monitoring and an increased understanding of the environmental problem, so that, if necessary, new targets and timetables can be implemented. Furthermore, if the minimum standard is set because one 'stakeholder' wants substantially lower standards, then if that 'stakeholder' is later shown to be wrong, they will be liable for penalties (or the other 'stakeholders' enjoy benefits) that are coupled to the opportunities lost by intensification of the problem and by lost time. The precise methodologies and mechanisms to be employed are still being developed (Swingland, in preparation).

7.6 EMPIRICAL EFFORTS TO RESEARCH AND RESOLVE THESE ISSUES

The derivation of policy and socio-economic insights to improve the costs and benefits of biodiversity management in tropical forests is relatively quick. It can be applied with comparative ease and some assurance of success. This is because researchers in these areas primarily use other people's data and their results can be used to adjust management practices directly. In stark contrast, science needs longer since it must first collect the field data, verify the validity of the information, analyse the results, and apply the results in the field. Relatively, the socio-economic and policy approach is faster, cheaper and more successful than science. Yet it is clear that both areas (physical and social sciences) must be effectively connected in a real, and quantified way, if biodiversity management (= conservation) and socio-economics are to be properly applied in long term development. It is necessary to encourage and promote dialogue between physical and social scientists.

Support for empirical research in the application of economic instruments needs to be accelerated, e.g. fiscal measures, such as taxes and subsidies, tradeable pollution permits, economic aspects of resource access rights, and valuation of common property resources. Further development of techniques of physical resource accounting and valuation should be done, with particular emphasis on valuing services and damages incurred through the use of environmental assets including transboundary externalities. Databases of relevant expertise in the field of environment and economics need refining. Promoting and applying methodologies to estimate the costs and benefits, and their distribution, of options arising in the negotiations on international environmental agreements is overdue. Developing economic and analytical frameworks for the institutional arrangements to address global environmental problems, including transfer payments and funding mechanisms, are badly needed by the planners. An analysis of linkage between international economic relations and environmental management is urgently needed, including: commodity terms of trade, relief and restructuring of external debts of developing countries, structural adjustment programmes, environmental standards as non-tariff trade barriers, conflict between trade and environment regulations/agreements, the conditionality of aid, impact of export subsidies and agricultural product subsidies as a means of improving the international base for environmental maintenance.

Scientists should continue to get better data to monitor the effects of 'stakeholders' and for all 'stakeholders' to understand better the nature and extent of the environmental problem, its long term consequences, and alternative technological solutions; and model, and use, data to examine how different environmental problems interact.

A particular problem in finding effective solutions to environmental issues is the difficulty of incorporating scientific results in effective policy. Scientists should therefore attempt to provide results in a form that is suitable for evaluating economic and social consequences. Analysis of socio-economic impacts of environmental change, particularly, linking information and data on the physical environment to those on human activities (e.g. linking geographical information systems to ecological and socio-economic data) are of particular significance.

It is clear that a number of institutions around the globe are trying to address these problems and provide sensible, useful solutions that can be incorporated in future development. Some of the more active groups are in fundamental species ecology and systematics, the National Museum of Natural History (Washington) and the Natural History Museum (London); in policy development, World Resources Institute (Washington); in data information and dissemination, World Conservation Monitoring Centre (Cambridge); in 'captive' plant conservation, Royal Botanic Gardens (Kew) and Missouri Botanic Gardens, USA; in captive animal conservation and re-introduction, Jersey Zoo (Jersey) and the National Zoo (Washington); in field programmes, Conservation International, Washington; in environmental economics, the London Environmental Economics Centre and

CSERGE, London; in conservation research and postgraduate education, The Durrell Institute (Canterbury, UK), the Institute of Zoology (London) and the University of Michigan (Ann Arbor, USA); and finally in tropical forests, the Oxford Forestry Institute, UK.

In my own Institute we are developing new conservation techniques by research and teaching tomorrow's leaders worldwide: the means of making tropical forest trees grow faster, and other species re-establish themselves in degraded desert, using mycorrhiza; the ability to weigh and monitor environmental change, improve planning, and restructure the internal organization of various companies for increased profit and environmental benefit (and train students in conservation business administration); to measure the effect of climate change on the genetics of amphibians, a key group; to provide insights into the genetics of threatened species; to re-model species management proposals; to train personnel and students from many countries in conservation through our unique postgraduate programme; to develop new computer design and graphic capabilities for information systems; to provide closed high-level seminars to improve communication and advance thinking; to support collaborative conserva-tion programmes with other institutions and countries involving both *in situ* and *ex situ* research and training, ensuring long term continuity and access to the most comprehensive and modern methods.

7.7 CONCLUSION

The new imperative of using and thereby saving tropical forests is at the forefront of the global agenda. The problems are political – whose forests are they?; economic – who's benefiting? who's paying?; and scientific. While the politico-economic wrangle will inevitably continue, it is the scientists who will be wrong-footed if they cannot be clear and precise in their advice and judgement when governments ask – what should we do and how? After all the physical scientists have been responsible for creating this new consciousness, this new imperative, and now they must deliver.

REFERENCES

Ackery, P.R. and Vane-Wright, R.I. (1984) *Milkweed Butterflies*, British Museum (Natural History), London.

Durbin, J. (in press) *The role of local people in the successful maintenance of protected areas in Madagascar*. Environmental Conservation.

Ehrlich, P.R. and Wilson, E.O. (1991) Biodiversity studies: science and policy. *Science* **253**, 758–62

Hawksworth, D.L. and Colwell, R.R. (eds.) (1992) Biodiversity amongst microorga-nisms and its significance. Biodiversity and Conservation 1:

Jenkins, M. (1992) Species extinction, in *Global Biodiversity* (ed. B. Groombridge), Chapman & Hall, London, pp. 192–205.

Laurance, W.F. (1991) Ecological correlates of extinction proneness in Australian tropical rain forest mammals. *Conservation Biology*, **5**, 79–89

Leemans, R. and Halpin, P.N. (1992) Global climate change, In *Global Biodiversity*, (ed. B. Groombridge), Chapman & Hall, London, pp. 254–5.

Maijala, K. (1990) Establishment of a world watch list for endangered livestock breeds, in *Animal Genetics Resources. A global programme for sustainable development*, (ed. G. Weiner), FAO Animal Production and Health Paper 80, FAO, Rome pp 167–84.

Mason, I.L. (1988) *A World Dictionary of Livestock Breeds, Types and Varieties*, 3rd ed. CAB International, Wallingford, UK.

May, R.M. and Southwood, T.R.E. (1990) Introduction, pl–22 in *Living in a Patchy Environment* (eds. B. Shorrocks and I.R. Swingland), Oxford University Press, UK, pp. 1–22.

Mittermeier, R.A. and Werner, T.B. (1990) Wealth of plants and animals unites 'megadiversity' countries. *Tropicus* **4** (1), 4–5

Myers, N. (1979)*The Sinking Ark: a new look at the problem of disappearing species*, Pergamon Press, Oxford, UK

Myers, N. (1990) The biodiversity challenge: expanded hot-spots analysis. *The Environmentalist*, **10**, 243–56

Poore, D., Burgess, P., Palmer, J. *et al.* (1989) *No Timber without Trees*, Earthscan, London.

Reid, W.V. (1992) How many species will there be? in *Tropical Deforestation and Species Extinction* (eds. T.C. Whitmore and J.A. Sayer), Chapman & Hall, London.

Reid, W.V. and Miller, K.R. (1989) *Keeping Options Alive: the Scientific Basis for Conserving Biodiversity*, World Resources Institute, Washington, USA.

Rojas, M. (1992) The species problem and conservation: what are we protecting? *Conservation Biology*, **6**, 170–8

Rosewell, J.P. (1990) Dynamic stability of a single-species population in a divided and ephemeral environment, in *Living in a Patchy Environment* (eds. B. Shorrocks and I.R. Swingland), Oxford University Press, pp. 63–74

Thomas, R. (1992) Genetic diversity, in *Global Biodiversity* (ed. B. Groombridge), Chapman & Hall, London, pp 1–6.

Thornback, J. and Jenkins, M. (1982) *The IUCN Mammal Red Data Book, Part 1*; IUCN, Gland, Switzerland and Cambridge, UK.

Vane-Wright, R.I., Humphries, C.J. and Williams, P.H. (1991) What to protect – systematics and the agonies of choice. *Biological Conservation*, **55**, 235–54.

Wright, S.J. (1990) Conservation in a variable environment: the optimal size of reserves, in *Living in a Patchy Environment* (eds. B. Shorrocks and I.R. Swingland). Oxford University Press, pp. 187–95.

Swingland, I.R. (in prep.) Weighing environmental stakeholder's stake: linking information and data on the physical environment to those on human activities.

8

Optimal economic growth and the conservation of biological diversity

Scott Barrett

8.1 INTRODUCTION

Conservationists are no longer concerned with the extinction of specific species alone, but with the loss of biological diversity generally. There are many reasons why diversity itself should be conserved, but perhaps the most important is the potential contribution of genetic information embodied in the species stock to basic research, particularly in medicine and agriculture.[1] Having recognized this potential, the conservation of biological diversity has become an important objective of public policy.[2]

The threat to biological diversity is not overexploitation (on which single

[1] See, for example, Harrington and Fisher (1982), Brown (1985) and Fisher and Krutilla (1985). Biological diversity is also believed to contribute to the stability of ecosystems, on which the earth's climatic and atmospheric conditions partly depend. Krutilla (1967) argues that biological diversity is valued for aesthetic reasons, and that an existence value exists for biological diversity. The environmental literature argues that biological conservation is a moral obligation.
[2] The World Bank (1986b, p. 26) has called the conservation of biological diversity 'an important Bank objective'. The International Environment Protection Act of 1983 made the conservation of species and their habitats 'an important objective of US development assistance' (U.S. Agency for International Development, 1985).

Economics and Ecology: New frontiers and sustainable development.
Edited by Edward B. Barbier. Published in 1993 by Chapman & Hall, 2–6 Boundary Row, London SE1 8HN. ISBN 0 412 48180 4.

species bioeconomic models have concentrated) but habitat destruction.[3] The preservation of natural environments like wildlife habitat was first posed as an economic problem in a seminal essay by Krutilla (1967). But while Krutilla's paper stimulated much research in this area,[4] the first item on Krutilla's (1967, p. 785) research agenda has not been explored previously:

> First, we need to consider what we need as a minimum reserve to avoid potentially grossly adverse consequences for human welfare. We may regard this as our scientific preserve of research materials required for advances in the life and earth sciences. While no careful evaluation of the size of this reserve has been undertaken by scientists, an educated guess has put the need in connection with terrestrial communities at about ten million acres for North America. Reservation of this amount of land – but a small fraction of one per cent of the total relevant area – is not likely to affect appreciably the supply or costs of material inputs to the manufacturing or agricultural sectors.

Of the world's many threatened ecosystems, tropical rain forests have the greatest potential for species extinction. While the rain forests cover only a small fraction of the earth's surface, they are believed to contain a majority of all wildlife species. Rates of deforestation have been so great that biologists worry that by the end of this century the only undisturbed tropical lowland forest left anywhere in the world may only be found in whatever wildlife reserves have been previously set aside.

The problem of conserving biological diversity in tropical rain forest countries is very different from the one described by Krutilla for North America. While the costs of preservation may not be very large in North America,[5] in poor countries preservation of natural environments is seen to be in conflict with development itself. Conservation biologists estimate that the minimum quantity of reserves needed to conserve just a majority

---

[3] It is interesting to note that while the literature has focused almost exclusively on problems of overexploitation, habitat destruction has caused almost all of the 500 extinctions known to have occurred in the US since European settlers first arrived. It is also interesting to note that two-thirds of these extinctions occurred in tropical Hawaii. See Harrington and Fisher (1982, p. 124). Miller (1981) has modelled the economics of habitat destruction involving a single species only. The question of extinction never arises in his paper, although conservation biologists attach great importance to the area requirements needed to support a minimum viable population of a species (see, for example, Frankel and Soulé (1981) and the contributions in Soulé (1986)).

[4] See, for example, the papers in Krutilla (1972) and Krutilla and Fisher (1985), Fisher, Krutilla and Cicchetti (1972), Krautkraemer (1985), Porter (1982) and Barrett (1992).

[5] According to section 7 of the U.S. Endangered Species Act of 1973, actions authorized, funded, or carried out by the federal government must not threaten endangered species or modify their critical habitat. Since most economic development projects having even modest environmental impact are required to obtain federal permits, section 7 provided a way to stop a great many projects that threatened endangered species, regardless of their social benefits. Harrington (1981), however, has found that the actual economic impact of the legislation was relatively small. This suggests that the costs of preservation are not very great.

of tropical species is 10% of the total area.[6] Furthermore, per capita incomes in tropical countries are very low (34 of the 36 so-called low income countries have tropical rain forests[7]), and these countries depend to a far greater extent on frontier development for their livelihood. Agriculture (including forestry) accounts for 36% of the GDP of low income economies but only 3% of the GDP of industrial market economies.[8]

The questions of how much virgin territory should be protected and at what rate the remaining forest area should be extracted are clearly linked to the larger question of development. This chapter develops an economic–ecological model, and solves for the optimal rate of deforestation and the optimal provision of wildlife reserves. The model takes well-being as depending on both (per capita) consumption and the stock of biological diversity that is conserved. The economic problem is to maximize the integral of discounted social utilities by choice of a rate of deforestation. The model assumes that the only way to conserve biological diversity is by preserving habitat – that is, by setting aside wildlife reserves.[9] It is further assumed that the appropriate measure of biological diversity is the number of species conserved.[10] Deforestation transforms the stock of virgin rain forests into a stock of developed natural capital, agricultural land, and in the process diminishes the stock of biological diversity. Consumption depends on the rate of deforestation directly (harvested timber) and on the income earned from the stock of developed capital (agricultural land).

Perhaps the most important lesson from the literature on the economics of environmental preservation is that changes in the underlying benefits and costs of frontier development can profoundly alter the optimal development policy when frontier development is irreversible. Fisher, Krutilla and Cicchetti (1972) prove that if preservation benefits are increasing relative to development benefits, then the optimal level of development will be less than an analysis based on current valuations would indicate is optimal. In other words, frontier development should cease at a time when the marginal benefit of development exceeds the marginal benefit of preservation.

---

[6] Myers (1986, p. 408). Myers does not state how he arrived at this figure. However, the figure can be obtained from the species–area relation described in the next section. If $\beta = 0.3$ then the preservation of 10% of habitat will conserve 50% of species.
[7] Tropical countries are identified in FAO (1982). Income classifications are from World Bank (1986a).
[8] World Bank (1986a). These statistics are likely to understate the case. In tropical countries there is a great deal of subsistence farming and logs are often felled illegally.
[9] Following Frankel and Soulé (1981, p. 4), 'conservation' is taken here to mean providing not only for the maintenance of species but also for their continuing evolution. By this definition, zoos and germplasm storage facilities preserve species; only wildlife reserves conserve species.
[10] Species diversity appears to be the most important criterion for choosing wildlife reserves. In a sample of 17 wildlife conservation studies published between 1971 and 1981, Usher (1986) found that all but one used diversity as a criterion. Species diversity can also be defined in terms of the distribution of the number of species according to their relative abundances. A superb analysis of the concept of 'diversity' is provided by Weitzman (1991).

In the case of deforestation, technical progress is likely to increase the consumption obtained both directly and indirectly from incremental deforestation. This would seem to work against the provision of wildlife reserves. However, technical progress that increases the returns to deforestation would also increase the productivity of the stock of developed capital, and growth in the consumption obtained from this capital stock may diminish the social value of additional consumption obtained by incremental deforestation. If the social value of the additional consumption from deforestation falls faster than the increase in consumption obtained by that deforestation, then preservation would appear attractive. Changes in the social value of additional consumption depend on the social welfare function that is employed, and hence on ethical views about the inter-generational distribution of consumption. Decisions about the provision of wildlife reserves therefore depend on ethical concerns as well as factual matters like the rates of technical progress and the important ecological relationships. This chapter makes clear the roles of ethical and factual parameters in development/conservation decisions. It shows that a strong case can be made that a greater area of virgin rain forest should be protected than would be suggested by an analysis of current costs and benefits.

The chapter is organized as follows. Section 8.2 discusses how species diversity depends on the area of habitat that is protected. Section 8.3 develops the economic–ecological model. Section 8.4 solves the model for the case where the underlying marginal benefits and costs of preservation do not depend on time. Section 8.5 solves the more difficult case where there is technical progress. The final section of the paper discusses the implications of the analysis for policy.

8.2 THE SPECIES–AREA RELATION

It is an empirical observation that the number of species (within a given taxonomic group) found in an area tends to increase with the size of the area.

Two hypotheses have been advanced to explain this. Williams (1964) among others, has argued that as the sample area is increased, the number of new habitats encountered should increase concomitantly. It follows that since new habitats harbour new species, the number of species found should depend positively on the size of the area sampled, all else being equal.

The second and more elegant hypothesis is really a prediction of island biogeography theory (see Preston (1962) and MacArthur and Wilson (1967)), which maintains that the number of species found within an area is determined by a dynamic balance between immigration and extinction rates. Consider an archipelago of islands. Suppose the species immigration rate for each island depends on the distance between the island and the mainland (the source of potential immigrants), but not on the size of the island. Suppose further that the species extinction rate for each island depends on the species population size, which in turn depends on the size of the island. Then if the distances between each island and the mainland were identical, the species extinction rate would be higher for small islands than large islands – that is, the number of species found would depend positively on island size. This hypothesis, like the first, extends naturally to wildlife reserves, which may be considered islands in a sea of altered habitat.

The species–area relation is usually expressed in the form

$$S = \alpha A^{\beta},\qquad\qquad (1)$$

where S is the number of species, A is area, and α and β are parameters.[11] The parameter α obviously depends on the units of area measurement. All else being equal, α will be larger the greater the species density. The parameter β is independent of the units of area measurement and can be interpreted as the elasticity of species diversity with respect to area. Estimates for β tend to fall in the range 0.18 – 0.35 (see Diamond and May, 1981). Thus the empirical evidence points to diminishing returns (in species diversity) with increasing area.[12]

I should emphasize that the species–area relation is static. Empirically it is estimated at a moment in time. According to island biogeography theory the relation would change if we were to observe S and A over time following some shock. If A were to fall, say, S would fall until the immigration and extinction rates were once again in balance. This process of 'relaxation' may take a very long time. Land bridge islands created by rising sea levels after the last ice age are a good laboratory for estimating relaxation time. Diamond (1973) has estimated that bird species on large land bridge islands off New Guinea in the Pacific have yet to reach a new equilibrium – 10 000 years after the islands were created. Small islands, however, reach a new equilibrium in much less time.[13]

An interesting example is Brown's (1971) study of mammals isolated on alpine 'islands' in the Great Basin Desert of North America. During the Pleistocene, pinon-juniper woodlands were driven down to elevations 2000 feet below their current levels (about 7500 feet). At this time species of mammals confined to pinon-juniper habitat were distributed contiguously

---

[11] The next most commonly estimated form is the semi-log (with S in levels regressed on A in logs plus a constant). See Connor and McCoy (1979) for a statistical and biological analysis of the species–area relation. See also Boecklen and Gotelli (1984) for cautionary remarks on the use of the species–area relation in directing conservation policies.

[12] In an analysis of 100 species–area relations, Connor and McCoy (1979) found only three with a $\beta < 0$ and two with a $\beta > 1$; the average was about 0.31.

[13] Another point to note is that in equilibrium, while the number of species in an area of given size will remain constant, the identities of the species may change significantly. Diamond (1973) found that between 17 and 62% of the bird species present 50 years ago on islands off New Guinea had disappeared and that an approximately equal number had immigrated.

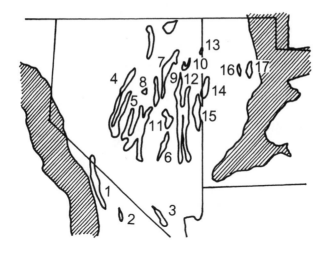

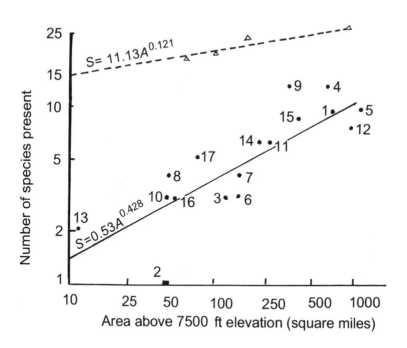

Figure 8.1 Estimation of the species–area relation. Source: Brown (1971) as reprinted in Frankel and Soulé (1981).

throughout the Great Basin. As temperatures rose about 8000 years ago, pinon-juniper habitat – and its associated mammal species – retreated to higher elevations. Seventeen 'habitat islands' were thus created in the Great Basin. These are shown in the map in Figure 8.1. The shaded area to the left is the Sierra Nevada mountain range, and the shaded area to the right is the Rocky mountains.

Brown identified 15 species of mammal found in the Sierra Nevada and Rocky Mountains and at least one of the 17 habitat islands of the Great Basin. He then regressed the number of these species found on each island on the island's size (both variables in logs). The result is the bottom regression in Figure 8.1. The top regression, based on only four observations, is for comparable habitat islands found within the Sierra Nevada. These four islands contain more species but have a smaller β coefficient. Brown attributes the high β estimate (0.43) for the Great Basin islands to an exceptionally low rate of colonization – lower than would be expected for oceanic islands.

The value of this information to biological conservation was first noticed in the mid-1970s (see May (1975) and Diamond (1976)). The immediate implication is that we can predict with some accuracy the number of species that will be conserved if a wildlife reserve of given size is set aside. For example, a β of about 0.3 implies that if 10% of the Amazonian rain forest were set aside as a wildlife reserve, roughly 25% of the original species would be conserved.

8.3 THE MODEL

Let S_t denote the stock of species diversity at time t, and let r_t denote the rate at which habitat area A_t is developed. The stock S_t is assumed not to be capable of regeneration; upon differentiating (1) with respect to time and substituting, the dynamics of the stock are described by

$$\dot{S}_t = -aS_t^{-b} r_t, \ S_0 > 0 \text{ given}, \ r_t \geq 0, \tag{2}$$

where $a = \beta\alpha^{1/\beta} > 0$, $b = -(\beta-1)/\beta$; $0 < \beta < 1$ implies $b > 0$.

Equation (2) was derived from (1), which is an equilibrium relation. Equation (2) thus tells us how the equilibrium value of S changes in response to a change in A. As noted earlier, the adjustment to a new equilibrium value of S (following a change in A) will not be instantaneous. In using (2), it is therefore assumed that our concern is with the number of species conserved permanently and not the number that are actually present at any particular moment in time.[14]

Consumption will generally depend not only on the rate of depletion, but also on the stock of the resource in its transformed or developed state.

[14] This accords with the biologist's definition of conservation (see footnote 9.)

Let D_t denote this stock. Then (2) implies, for $t \geq 0$,

$$D_t = D_0 + \int_0^t r_\tau \, d\tau,$$

assuming the total quantity of resources is fixed; i.e., $D_t - D_0 = A_0 - A_t$. If A is expressed as hectares of virgin rain forest, D would be measured in hectares of managed forest or agricultural land. Notice that while the virgin rain forest is assumed not to be capable of regeneration, forests do grow on the managed plantation, although the details of the optimal forest rotation are subsumed within the model.

Let the production function for the developed sector of the economy be given by $F(D_t)$, and assume $F_D > 0$ and $F_{DD} < 0$. It will simplify matters if F is expressed in terms of the original state variable, S. Since $A_0 + D_0$ is fixed, we can write $F(D_t) = F(A_0 + D_0 - \alpha^{1/\beta} S_t^{1/\beta}) = f(S_t)$. It is easy to verify that $f_s < 0$ and $f_{ss} < 0$ for $\beta < 1$.

Assuming that all output is consumed, total consumption C_t can be written

$$C_t = \sigma e^{\gamma t} r_t + f(S_t) e^{\omega t}, \tag{3}$$

where γ and ω are the (assumed constant, but not necessarily positive) rates of technical progress in the frontier and developed sectors, respectively, and σ is a constant that converts the rate of depletion in the initial period into a consumption rate. Throughout this chapter it is assumed that $\sigma > 0$ and $f(S_t) > 0$ for all S_t.[15]

If we assume zero population growth and normalize by setting population at $t = 0$ equal to one, then C_t can be interpreted either in total or per capita terms. We can then write the instantaneous social utility function as $U(C_t, S_t)$. Assume $U_c, U_s > 0$; $U_{cc}, U_{ss} < 0$; $U_{sc} = U_{cs} = 0 \ \forall C_t, S_t$; and $U_c(0) = \infty$. Further, let the elasticity of the marginal social utility of consumption, $\eta = -U_{cc}C/U_c$, be a positive constant.[16]

The economy's problem is to

$$\max_{\{r_t\}} \int_0^\infty U(C_t, S_t) e^{-\delta t} \, dt, \tag{4}$$

for $\delta > 0$, subject to (2) and (3). The current value Hamiltonian is

$$H = U(\sigma e^{\gamma t} r_t + f(S_t) e^{\omega t}, S_t) - \lambda_t r_t a S_t^{-b}, \tag{5}$$

---

[15] These assumptions are weakened in Barrett (1992).

[16] For example, we might have $U(C,S) = V(C) + W(S)$ with $V(C) = -C^{-(\eta-1)}/(\eta-1)$ for $\eta > 0$, $\eta \neq 1$ and $V(C) = \log C$ for $\eta = 1$. The constant elasticity assumption is common in the literature and is especially helpful in simplifying the analysis in Section 4. Barrett (1992) also considers the case where $\eta = 0$.

and the first order conditions are (2) and

$$r_t \geq 0 \text{ if } U_c(\sigma e^{\gamma t} r_t + f(S_t)e^{\omega t})\sigma e^{\gamma t}/aS_t^{-b} = \lambda_t \qquad (6a)$$

$$r_t = 0 \text{ if } U_c(f(S_t)e^{\omega t})\sigma e^{\gamma t}/aS_t^{-b} \leq \lambda_t \qquad (6b)$$

$$\dot{\lambda}_t = \delta \lambda_t - U_s(S_t) - U_c(\sigma e^{\gamma t} r_t + f(S_t)e^{\omega t})f_s(S_t)e^{\omega t} - \lambda_t r_t abS_t^{-b-1}. \qquad (7)$$

Since $H_{ss} < 0$ and $H_{rr} < 0$ for $U_{cc} < 0$, the transversality conditions,

$$\lim e^{-\delta t}\lambda_t S_t = 0, \lim e^{-\delta t}\lambda_t \geq 0 \qquad (8)$$

$$t \to \infty \qquad\qquad t \to \infty$$

are sufficient for a maximum.

8.4 SOLUTION TO THE AUTONOMOUS CONTROL PROBLEM

It will prove useful to begin by solving the autonomous problem where $\gamma = \omega = 0$.[17] Assuming an interior solution, (6a) may be differentiated with respect to time to yield, after substitution,

$$C_t/C_t = [U_s(S_t)aS_t^{-b}/\sigma U_c(f(S_t) + \sigma r_t) + f_s(S_t)aS_t^{-b}/\sigma - \delta]/\eta. \qquad (9)$$

The steady state value for $\tilde{S}$, is defined by

$$U_s(\tilde{S})/\delta = \sigma U_c(f(\tilde{S}))/a\tilde{S}^{-b} - U_c(f(\tilde{S}))f_s(\tilde{S})/\delta. \qquad (10)$$

The LHS of (10) is the marginal benefit of species preservation. The RHS is the marginal benefit of deforestation, and is composed of two terms. The first is the value of the consumption obtained by the incremental deforestation at the time that deforestation occurs. The second is the value of the consumption obtained by increasing the stock of agricultural land by one (species) unit. It is easy to verify that the LHS of (10) is decreasing in S, and that the RHS is increasing in S. Hence, the steady state is unique.

Define the quantity of wildlife reserves as the quantity of virgin territory left undeveloped in the steady state. It is straightforward to prove

Proposition 1. If $\gamma = \omega = 0$, then the optimal rate of deforestation is always positive but declining, and it is optimal to set aside some positive quantity of wildlife reserves.

The phase diagram for this problem is shown in Figure 8.2.

---

[17] The problem is autonomous insofar as time enters problem (4) only through the positive rate of pure time preference.

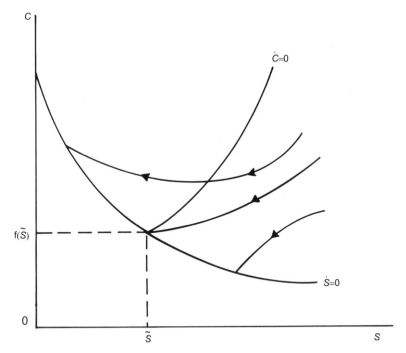

Figure 8.2 Phase diagram for autonomous control problem.

8.5 SOLUTION TO THE NONAUTONOMOUS CONTROL PROBLEM

The previous section showed that if the economic relations in the problem do not depend on time, then some permanent preservation of virgin territory is optimal. The reason, of course, is that with $f > 0$, the economy can meet its basic needs without always having to deplete its rain forests. Krutilla's (1967) seminal paper, however, warns us to be alert to the possibility that the economic relations in the problem may vary over time. This section considers the more general problem where $\gamma, \omega \neq 0$. As indicated in the introduction, it is this problem that is most important to the analysis of deforestation and the provision of wildlife reserves.

From an analytical point of view, one can distinguish between cases where optimal deforestation ceases in finite time and cases where optimal deforestation ceases only in the limit. This chapter considers in detail only the former cases. If $S_0 > \tilde{S}$, then a sufficient condition for optimal frontier development to cease in finite time is that the marginal benefit of frontier development be decreasing over time for $r_t = 0$. The following lemma states the conditions that are sufficient (but not necessary) for the marginal benefit of deforestation to be decreasing over time when $r_t = 0$.

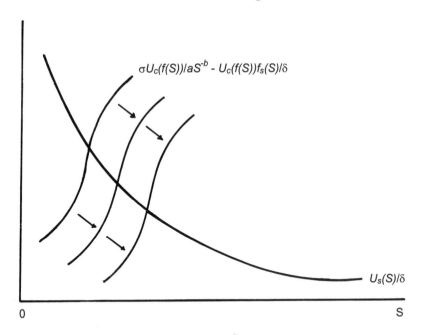

$$\sigma U_c(f(S))/aS^{-b} - U_c(f(S))f_s(S)/\delta$$

$$U_s(S)/\delta$$

0   S

Figure 8.3 Illustration of changes in marginal relationships over time (with $r_t \geq 0$, $\sigma(\eta\omega-\delta \geq 0$, $\omega(\eta - 1) \geq 0$ and at least one of these inequalities holds strictly).

Lemma. The marginal benefit of deforestation is decreasing for $r_t = 0$ if $\sigma(\eta\omega-\gamma) \geq 0$ and $\omega(\eta-1) \geq 0$, and at least one of these inequalities holds strictly.

Proof. The lemma is best proved informally. The marginal benefit of deforestation (assuming $r_t = 0$) is

$$\sigma e^{\gamma t} U_c(f(S)e^{\omega t}/aS^{-b} - U_c(f(S)e^{\omega t})f_s(S)e^{\omega t}/\delta. \tag{11}$$

Consider the first term in (11). This term rises at rate γ, holding U_c fixed. For the entire term to be nonincreasing, U_c must therefore fall at a rate at least as great as γ. U_c falls at a rate η times the rate of increase of C_t when $r_t = 0$, or $\eta\omega$. Hence the first term in (11) is nonincreasing if and only if $\eta\omega \geq \gamma$, and decreasing if this inequality holds strictly. Now consider the second term in (11). We know that U_c falls at rate $\eta\omega$, and that $-f_s e^{\omega t}$ rises at rate ω. The entire second term is therefore nonincreasing if and only if $\eta\omega \geq \omega$, or $\omega(\eta-1) \geq 0$, and decreasing if this inequality holds strictly. If both of these inequalities hold, then the marginal benefit of deforestation is nonincreasing; if at least one inequality holds strictly, then the marginal benefit of deforestation is decreasing over time for $r_t = 0$. This last case is illustrated in Figure 8.3.

Since the marginal benefit of preservation is stationary for $r_t = 0$, if the marginal benefit of deforestation is decreasing, it must be optimal to cease frontier development at some finite time T and to preserve some positive S. Proposition 2 establishes these values.

Proposition 2. If $\omega(\eta\text{-}1) \geq 0$, and at least one of these inequalities holds strictly, then it is optimal to develop the rain forest (i.e. choose $r_t > 0$) on the interval $(0, T)$, and to cease deforestation (i.e. choose $r_t = 0$) at time T and stock $S_T = \tilde{S}$ where I and $\tilde{S}$ satisfy

$$U_s(\bar{S})/\delta = U_c(f(\bar{S})e^{\omega T}\sigma e^{\gamma T}/a\bar{S}_T^{-b} - U_c(f(\bar{S})e^{\omega T})f_s(\bar{S})e^{\omega T}/[\omega(\eta-1)+\delta]. \quad (12)$$

Proof. See Barrett (1992).

Equation (12) contains two unknowns. However, $\tilde{S}$ and T and can be solved using the transversality conditions for the finite horizon formulation of (4); see Barrett (1992).

Proposition 2 tells us that if decisions regarding the provision of wildlife reserves were made on the basis of only current values for costs and benefits, too little virgin territory would be protected if $\omega(\eta\text{ -}1) = 0$ and $\gamma(\eta\omega - 1) \geq 0$, and at least one of these inequalities held strictly.

In the case of deforestation, ω would represent the rate of technical progress in agriculture or plantation forestry, and γ would represent the rate of technical progress in harvesting primary forests. If these two parameters are positive, the returns to deforestation would be increasing over time. This would seem to make deforestation appear ever more attractive over time. However, technical progress in agriculture would increase the consumption obtained from the stock of developed capital – agricultural land – and not just the return to frontier development. If the consumption obtained from the stock of already developed capital increases over time, then development of the frontier can appear less attractive. Whether this is the case depends on the parameter η.

The parameter η is the elasticity of the social marginal utility of consumption; it is thus an ethical parameter. The larger is η, the greater is the concern for a more 'egalitarian' distribution of consumption across generations.[18] This can be seen by referring back to the autonomous control problem. Equation (9) shows that the larger is η, the smaller is the rate of change in consumption. When η is large, society does not want one generation to be made substantially better off at the expense of another. With $\omega > 0$, future generations will have higher consumption levels, even if the environment is not developed. Under these circumstances, the larger is η, the more attractive preservation appears. Indeed, in the limit as $\eta \to \infty$ one obtains the maxi-min solution; all remaining rain forests are preserved.

The other ethical parameter, δ, discriminates only against time. The larger is δ, the less important is the well-being of future generations relative to the present. In a sense, then, δ and η work in opposite directions. This can be seen from Equations (9) and (12). Equation (9) shows that the larger

[18] See Dasgupta and Heal (1979), especially Chapter 10.

is η, the more even is the distribution of consumption across generations; the larger is δ, the more consumption is tilted to the present. As η is increased in Equation (12), the marginal benefit of deforestation declines; as δ is increased, the marginal benefit of deforestation rises relative to the marginal benefit of preservation. Importantly, while δ helps determine the steady state, and the optimal path leading to the steady state, it does not influence the decision of whether deforestation should cease at a time when the marginal benefit of preservation is less than the associated marginal cost.

The ecological parameter β also determines the optimal provision of wildlife reserves but does not affect the decision of whether deforestation should cease when the marginal benefit of preservation is less than the associated marginal cost. We can see from (12) that the larger is the parameter a and the smaller is the parameter b, the smaller is the marginal benefit of frontier development. The parameter a increases with β, while the parameter b decreases with β. Hence, the larger is β, the larger is the area that should be set aside as a wildlife reserve.

We see then that the fate of the rain forests should depend on both positive and normative parameters. The rates of technical progress in frontier development (γ and ω) are empirical parameters. However, unlike the ecological parameter β, the values taken by γ and ω will depend, among other things, on public sector investment in R&D and infrastructure, and on pricing policies, and so are under the influence of governments and development agencies. If it were believed that R&D was less than optimal, then one might be tempted not to use the observed values for γ and ω, but to substitute values for γ and ω that were consistent with optimal R&D. However, one should employ values for γ and ω based on what one actually expects them to be, and not on what one thinks they should be. One may believe that ω is 'too low' and γ 'too high', but to substitute one's preferred values for ω and γ will mean that social welfare is not actually maximized. At the same time, decisions about public sector policies and investments that influence ω and γ should take into account the effect these policies and investments have on development/preservation of the environment. In other words, the issue for development programmes is not just how much growth but what kind of growth should be provided. Increasing the productivity of the developed sector makes greater environmental preservation optimal, and this in turn can increase general well-being.

Estimation of the two ethical parameters, η and δ, is probably not appropriate. One can observe how a country makes decisions that affect income distribution – like tax policy – and infer from these observations a value for η.[19] Similarly, one can observe how society makes trade-offs between consumption now and in the future, and infer values for δ. However, to argue that the estimated values actually represent η and δ is to impute optimality to the observations. What is more, since more is taken into account in deciding on such policies than these parameters, estimation

[19] See, eg. Stern (1977), who estimates values for η ranging from one to three.

of the ethical parameters from observed data is virtually impossible. As Dasgupta, Marglin and Sen (1972, p. 120) argue, parameters like η and δ 'should reflect conscious political decisions with respect to what are, after all, political questions'.[20]

8.6 CONCLUSION

The economic–ecological model presented in this chapter finds that the conservation of biological diversity depends as much on society's ethical views as on facts. Even where different parties agree on the facts, they may disagree on the ethical parameters. Economics and ecology cannot say how much virgin forest should be protected from development, but they can provide a framework for forming the answer to this question, and for distinguishing between the normative and positive elements of the answer.

The most important conclusion reached by this chapter is that, if certain conditions hold, more habitat should be protected than would be indicated by a comparison of current values for costs and benefits. This is not because uncertainty creates option value or quasi-option value; the model presented in this chapter assumes complete certainty. The conclusion follows from the fact that loss of biological diversity is irreversible, and that the marginal benefit of preservation rises over time relative to the marginal costs if these conditions hold. 'Too much' biological diversity is conserved initially, and 'too little' is conserved later. However, just the 'right' amount of biological diversity is conserved on average.

The conditions that must hold to obtain this outcome are that $\omega(\eta - 1) \geq 0$ and $\gamma(\eta\omega - 1) \geq 0$, and at least one of these inequalities holds strictly. However, these conditions emerge from a rather specialized model. If one adds a non-resource sector, then the case for preservation is again likely to be supported. Suppose $F = 0$ but that consumption in the absence of frontier development (that is, when $r_t = 0$) is positive but grows at rate κ. Then it is easy to show that $\eta\kappa > \gamma$ implies that it will be optimal to cease frontier development in finite time, and to protect more remaining natural environments than is indicated by current valuations. Similarly, Krautkraemer (1985) shows that if the resource is combined with a stock of human-made capital to produce an income stream, then permanent preservation can be optimal if the (assumed constant) elasticity of substitution between capital and the resource exceeds the inverse of the elasticity of the marginal utility of consumption (η).[21] This last finding is similar to the one demonstrated in this chapter. Here, development of the frontier creates a form of natural

---

[20] At the same time, Koopmans (1965) cautions against using parameter values solely on the basis of their ethical merits. It is important also that the parameter values be chosen to ensure that an optimal programme exists, and that the optimal programme itself accords with ethical thought.

[21] One also requires that the marginal productivity of capital exceed δ as the capital stock becomes infinitely large, and that the initial capital stock be 'large'. See Krautkraemer's second model.

capital which raises consumption and thereby eases the need to develop the frontier further. In Krautkraemer's model, human-made capital is substituted for the resource input, enabling consumption to rise even while the rate of extraction is progressively lowered. All of these results hold when preservation benefits do not depend on time. If preservation benefits are expected to increase over time for S and C given, perhaps because of a change in preferences, then the analysis presented here would be reinforced; the case for preservation would be strengthened further.

Pulling all these results together, one is able to build a strong case for conserving more biological diversity than indicated by current valuations. This case depends on ethical as well as economic and ecological parameters – a result that I find reassuring, given that the analysis is intended to help decide the fate of environments such as the Amazonian rain forests, as well as the course of development itself.

ACKNOWLEDGEMENTS

This chapter provides an application of the theoretical apparatus presented in Barrett (1992). I would like to thank Partha Dasgupta, Jeffrey Krautkraemer, Karl-Göran Mäler, David Pearce and Robert Solow for commenting on earlier versions of this work.

REFERENCES

Barrett, S. (1992) Economic growth and environmental preservation, *Journal of Environmental Economics and Management*, forthcoming.

Boecklen, W. and Gotelli, N.J. (1984) Island biogeographic theory and conservation practice: species–area or specious–area relationships? *Biological Conservation*, **29**, 63–80.

Brown, G. (1985), 'Valuation of Genetic Resources,' paper prepared for the Workshop on Conservation of Genetic Resources, 12–16 June, Lake Wilderness, WA.

Brown, J.H. (1971) Mammals on mountaintops: nonequilibrium insular biogeography, *American Naturalist*, **105**, 467–78.

Connor, E.F. and McCoy, E.D. (1979) The statistics and biology of the species–area relationship, *American Naturalist*, **113**, 791–833.

Dasgupta, P.S. and Heal, G.M. (1979) *Economic Theory and Exhaustible Resources*, Cambridge University Press, Cambridge.

Dasgupta, P., Marglin, S. and Sen, A. (1972) *Guidelines for Project Evaluation*, New York: UNIDO.

Diamond, J.M. (1973) Distributional ecology of New Guinea birds, *Science*, **179**, 759–69.

Diamond, J.M. (1976) Island biogeography and conservation: strategy and limitations, *Science*, **193**, 1027–9.

Diamond, J.M. and May, R.M. (1981) Island biogeography and the design of nature reserves, in *Theoretical Ecology: Principles and Applications*, 2nd ed., (ed. R.M. May), Blackwell, Oxford pp. 228–52.

Fisher, A.C. and Krutilla, J.V. (1985) Economics of nature preservation, in *Handbook of Natural Resource and Energy Economics*, vol. 1, (eds A.V. Kneese and J.L. Sweeney), Elsevier, Amsterdam; pp. 165–89.

Fisher, A.C., Krutilla, J.V. and Cicchetti, C.J. (1972) The economics of environmental preservation: A theoretical and empirical analysis, *American Economic Review*, **64**, 1030–9.

Food and Agriculture Organization (1982) *Tropical Forest Resources*, FAO, Rome.

Frankel, O.H. and Soulé, M.E. (1981) *Conservation and Evolution*, Cambridge University Press, Cambridge.

Harrington, W. (1981) The endangered species act and the search for balance, *Natural Resources Journal*, **21**, 71–92.

Harrington, W. and Fisher A.C. (1982) Endangered species, in *Current Issues in Natural Resource Policy*, (ed. P.R. Portney), Resources for the Future, Washington, DC, pp. 117–48.

Koopmans, T.C. (1965) On the concept of optimal economic growth, *Pontificiae Academiae Scientiarum Scripta Varia*.

Krautkraemer, J.A. (1985) Optimal growth, resource amenities and the preservation of natural environments, *Review of Economic Studies*, **52**, 153–70.

Krutilla, J.V. (1967) Conservation reconsidered, *American Economic Review*, **57**, 777–86.

Krutilla, J.V. (ed.) (1972) *Natural Environments: Studies in Theoretical and Applied Analysis*, Johns Hopkins University Press, Baltimore.

Krutilla, J.V. and Fisher, A.C. (1985) *Natural Environments: Studies in Theoretical and Applied Analysis*, 2nd ed., Resources for the Future, Washington, DC.

MacArthur, R.H. and Wilson, E.O. (1967) *The Theory of Island Biogeography*, Princeton University Press, Princeton.

May, R.M. (1975) Island biogeography and design of wildlife reserves, *Nature*, **254**, 177–8.

Miller, J.R. (1981) Irreversible land use and the preservation of endangered species, *Journal of Environmental Economics and Management*, **8**, 19–26.

Myers, N. (1986) Tropical deforestation and a mega-extinction spasm, in *Conservation Biology: The Science of Scarcity and Diversity*, (ed. M.E. Soulé), Sinauer Associates, Sunderland, Mass.

Porter, R.C. (1982) The new approach to wilderness preservation through benefit-cost analysis, *Journal of Environmental Economics and Management*, **9**, 59–80.

Preston, F.W. (1962) The canonical distribution of commonness and rarity, *Ecology*, **43**, 185–215 and 410–32.

Soulé, M.E. (ed.) (1986) *Conservation Biology: The Science of Scarcity and Diversity*, Sinauer Associates, Sunderland, Mass.

Stern, N. (1977) Welfare weights and the elasticity of the marginal valuation of income, in *Studies in Modern Economic Analysis*, (eds. M. Artis and R. Nobay), Blackwell, Oxford.

US, Agency for International Development (1985), *U.S. Strategy on the Conservation of Biological Diversity: An Interagency Task Force Report for Congress*, Washington, DC: USAID, Washington DC.

Usher, M.B. (1986) Wildlife conservation evaluation: attributes, criteria and values, in Wildlife Conservation Evaluation, (ed. M.B. Usher), Chapman & Hall, London.

Weitzman, M.L. (1991) On diversity, mimeo, Department of Economics, Harvard University.

Williams, C.B. (1964) *Patterns in the Balance of Nature*, Academic Press, London.

World Bank (1986a) *World Development Report 1986*, Oxford University Press, New York.

World Bank (1986b) *World Bank Annual Report 1986*, World Bank, Washington, DC.

9

The viewing value of elephants

Gardner Brown, Jr.
With the assistance of Wes Henry

9.1 INTRODUCTION

One way to marshal the resources necessary to reduce poaching effectively is to demonstrate the opportunity cost or economic loss associated with declines in elephant populations. It is this task which unites the concerns of environmentalists and naturalists with the subject matter of economists which we now address.

Elephants have many valuable characteristics. Their ivory can be harvested and grows more valuable with the age of the elephant. Many would like to see the elephants preserved as a species either for their own sake or their availability to be viewed by others. These two sources of economic value are important but are not the concern of this chapter which is more limited: to estimate the viewing value of elephants.

Viewing elephants creates two types of economic benefits. First, tourists spend money on their safaris, providing employment and profits to Africans and non-Africans. Second, tourists may receive consumers' surplus, meaning that they obtain more satisfaction, or value their safari more than it costs. From an international perspective, both types of value are important. However, African countries may be less concerned with calculations of consumers' surplus since it is incident on foreigners. The net economic value, or consumers' surplus, from an international perspective, is the subject treated in the following sections. Estimated values are obtained using two standard procedures in recreation economics: travel cost and contingent valuation in that order. Tourist expenditures are discussed briefly.

Economics and Ecology: New frontiers and sustainable development.
Edited by Edward B. Barbier. Published in 1993 by Chapman & Hall, 2-6 Boundary Row, London SE1 8HN. ISBN 0 412 48180 4.

The pragmatic objective was to estimate the total net economic value of a safari and then to allocate some plausible portion of that total to elephants since a proper marginal analysis was not feasible. The allocation exercise is straightforward. Estimating the demand function for safaris, from which consumer surplus is calculated, was more difficult.

9.2 VIEWING VALUE OF ELEPHANTS[1]

9.2.1 Travel cost method

A standard way to estimate the demand for a recreation activity is a travel cost procedure conceived by Hotelling and implemented by Clawson and Knetch (1966). It exploits the fact that people coming from different locations, countries in this instance, pay a different price to experience the same quality safari because travel costs differ. Estimating the relationship between different rates of participation at different prices, controlling for factors other than price such as income in this estimation, yields a demand function. Consumer surplus (CS) is the difference between what some people pay, say $OPBC$ in Figure 9.1, and the maximum they would pay ($OABC$). Thus PAB is the net economic value of a safari. This value neither

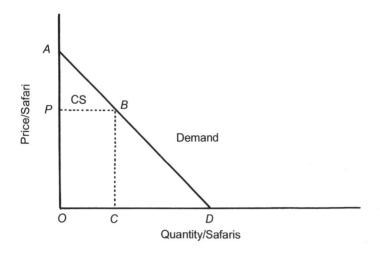

Figure 9.1 Demand and consumers' surplus for safaris.

[1] Neal Johnson and David Colpo assisted in the research process. Without the able work of Dixon Sarikoki, this project would not have been completed. Various conservation organizations and their employees were especially helpful at critical moments, including Holly Dublin and Helen deButts. Dr Richard Leakey, Director of the Wildlife Conservation and Management Department kindly granted permission to distribute the survey to tourists. Daniel Sindiyo, Director of the Kenya Tour Operators Association was helpful in the tour operators survey.

Table 9.1 Price data (US dollars)

Region	Land cost	Air fare	Travel time	Hourly wage	Weight	Time cost	Total price
N. America	1465	1900	40 h.	22.50	0.30	270	3635
Europe	957	1300	18 h.	22.50	0.30	121	2378

shows up in market observations nor is it captured by tour operators or trinket sellers. Yet it is a measurable, real economic loss international society would bear if safaris were prohibited. In contrast, if safaris stopped, people would spend their money elsewhere, so the forgone expenditures are not an economic loss, again from an international perspective.

Ideally, we would recognize explicitly that there are many qualities of safaris and estimate a set of simultaneous equations, one for each quality of safari. A non-random and inadequate sample size were sufficient reasons to search for a simpler solution.[2,3]

Since about 80% of the tourists to Kenya came either from North America or Europe,[4] participation figures from Europe and North America were used to construct a linear demand curve from these two points. The other 20% of the tourists were assumed to have the same average consumers' surplus as the included 80% have.

Quantity

The quantity variables for North America and Europe were derived from the *Economic Survey* for Kenya (1989). These included visitors on holiday in Kenya. There were 63 000 visitors from North America and 350 000 from Europe or 0.2316 per 1000 population for North America and 0.9826 for Europe, using population estimates from North America and Western Europe for 1988.[5]

Price ($US)

Price is defined as the sum of land costs, air fare and travel time costs. See Table 9.1 for a summary.

[2] Those knowledgeable about the travel cost method will recognize the delicate problems here of normalizing and sample size under conditions of constant quality. Into how many zones should the US be divided?

[3] Since Europeans do not travel as far as North Americans, the cost or price of any given quality safari differs between these two regions. Observable price differences for the same good at a moment in time permits the estimation of demand function or functions, if only the right data can be collected. Experiments with a multinominal logit function produce unreliable results because a very small fraction of the actual sample came from Europe.

[4] About 2% came from Africa and the rest from other countries, according to the *Statistical Abstracts* for Kenya for recent years.

[5] The countries included are: Austria, Belgium, Denmark, Finland, France, Greece, Ireland, Italy, Luxembourg, Netherlands, Norway, Portugal, Spain, Sweden, Switzerland, UK and West Germany.

Land costs were estimated by creating a quality weighted price index from the tour operators' surveys.[6] For example, the land cost faced by North American visitors, P_n, is defined as

$$P_n = \sum_j \left(\frac{P_{nj} * N_{nj}}{C_n}\right) * P_{nj}$$

where

j = tour operator;
C_n = total expenditures of North America;
N_{nj} = number of North American visitors tour operator j served in the last 12 months
P_{nj} = tour price of average length safari offered by tour operator j.

Air cost for a North American was $1900, the average air fare from the 20 respondents who indicated a North American residence and answered the air fare question. There were only 4 responses from Europe which gave air fares ($767, $1600, $1600, $5000) and $1300 was selected as a representative round trip air fare for Europe.

When the opportunity cost of travel time was included, travel time costs were estimated as the product of hourly wage, round trip travel time and a 30% weighting to reflect the common view that the value of time, net of taxes, while going on vacation is less than the gross wage rate. From the visitors' surveys, average income for both North America and Europe was estimated as $45 000. It was further assumed that visitors worked an average of 250 eight-hour days each year, yielding an hourly wage of $22.50. Travel time was also estimated from the visitors' surveys. Round trip travel time was 40 hours for North America and 18 hours for Europe.

The estimated demand function

Table 9.2 summarizes the price and quantity variables derived above. Fitting a straight line demand relation through these two points gives the following inverse relation:
$$P = 4023 - 1674Q$$
where P is the sum of land, air and travel time costs, and Q is holiday visits per 1000 population. Note that we have yet to address the problem of the percentage of visitors on holiday that actually goes on safari.

Table 9.2 Summary of price and quantity

Region	Price (US $)	Quantity (visits per 1000)
North America	3635	0.2316
Europe	2378	0.9826

[6] With hundreds of tour operators serving the market, it is unlikely that price discrimination is widespread.

Given the linear demand curve, per person consumer surplus for North America is $194 and for Europe it is $822.50.

It seems reasonable that a safari, a once-in-a-lifetime adventure for most North Americans, most of whom reportedly had a very satisfactory experience, would be worth 5% more than the cost. It seems plausible that a similar experience at less cost would be worth 35% more than the cost for a European. We suspect that if there is a bias, it has understated the North American and overstated the European value. Average consumers' surplus, estimated by weighting the above estimates by the number of holiday visitors for North America and Europe, is $726.63.

Total consumers' surplus for all those on safari requires an estimate of the total number of people who visited parks and reserves and paid the prices used in the above computations. The number of people is not recorded. Many holiday visitors go to the coast of Kenya and don't visit parks or reserves. The Kenya *Economic Survey* (1989) and the *Statistical Abstracts* for various years contain estimates of the number of visitors to parks and reserves but there are no formal estimates of the number of parks visited by the average tourist on safari so the gross data are not useful. Visitor data do not distinguish between children and adults.

Our estimate of between 250 000 and 300 000 adults per year who went on safari in 1988 is based on discussions with tour operators and with personnel in the economic section of the US Embassy in Kenya. Using these values, estimated total consumer surplus for those on safari varies between $182 and $218 million annually, depending on the assumed level of visitation.

Elephant's share of net economic value

Obviously not all of the value of a safari can be attributed to elephants because a safari is a bundle of satisfactory experiences. Tourists on safari were asked to allocate the pleasure and enjoyment of their trip over four stipulated categories of experience, including 'seeing, photographing and learning about the wildlife'. This element earned about 50%.[7]

Respondents were then asked to apportion their enjoyment of wildlife across four specified categories including 'seeing African elephants', which

[7] People travel to East Africa for many reasons. Thinking about the pleasure and enjoyment you are experiencing (or have experienced) from your visit, what percent of your pleasure would you attribute to each of the following? (*Please make your responses add to 100 percent*)

	Percent	
	Mean	Median
Seeing, photographing and learning about the wildlife	50	50
Accommodations, staff and services, drivers	20	20
Observing and learning about Africa and its cultures	19	20
Rest, relaxation, and shopping	9	10
Other experiences	2	0
	100%	100%

Table 9.3 Estimated annual viewing value of elephants (millions US dollars)

Cases	Travel cost method	Contingent valuation method
Value of time = 0		
Low visitor estimate	20	
High visitor estimate	24	
Positive value of time		
Low visitor estimate	23	
High visitor estimate	27	
Average willingness to pay		
Low visitor estimate		22
High visitor estimate		27
Median willingness to pay		
Low visitor estimate		25
High visitor estimate		30
Best point estimate	25	

attracted an average value of 25%.[8] The results of these two questions indicate that viewing elephants is the source of 12.6% of the total value of a trip.

Applying the share of 12.6% attributed by the visitors to elephants to the estimated economic value of a safari yields a viewing value for elephants of $23 to $27 million per year based on the travel cost method. See Table 9.3.

While it is routine in recreation economics to include some value for the opportunity cost of time, non-economists may be interested in its impact on the estimated values. When the value of time is given a zero value, the viewing value of elephants is reduced modestly by about 12% to between $20 and $24 million per year (Table 9.3).

9.3 CONTINGENT VALUATION METHOD

The second way in which the viewing value of elephants is measured is

[8] Thinking just about the *wildlife* and the pleasure and enjoyment it has or is giving you, what percent of your enjoyment of the wildlife would you attribute to each of the following? (*Please make sure your responses add to 100 percent*)

	Percent	
	Mean	Median
Seeing the big cats including lion, leopard, and cheetah	28	30
Seeing large numbers of a variety of wildlife species	29	25
Seeing African elephants	25	20
Learning about the ecology and animal behavior	16	15
Other (specify):	2	0
	100%	100%

contingent valuation. For more than two decades, economists have been conducting surveys in which sample populations are asked hypothetical questions designed to elicit how much money they are willing to pay or to receive in exchange for the preservation, maintenance, increase or decrease of some amount of a natural resource quantity or quality, for some specified amount of time. The format of these contingent valuation questions varies, depending on the nature of the resource and the policy question at hand. The economics profession is not of one mind about the accuracy of answers to contingent valuation questions. Two surveys of this literature are Mitchell and Carson (1989) and Cummings, Brookshire and Schulze (1986).

All agree that the validity of the answers depends on the care with which questions are designed in order to remove a variety of biases. Practically speaking, there are some crucially important natural resource policy issues which cannot be addressed unless contingent valuation questions are asked, for example, when there are no actual or possible observable data regarding behavioural responses to changes in natural resource availability. No observable behaviour will reveal accurately what people are willing to pay to keep elephants at their current population levels in any given region of Africa. The necessity to use contingent valuation procedures to address critically important policy issues has been recognized explicitly in the United States by the Water Resources Council (1983) and by Congress in specifying how damages from oil spills and hazardous waste are to be estimated.[9]

The tourists' survey contains a series of contingent valuation questions. Illustratively, people asked to pay for a special annual permit (or increased safari cost) of $100 which would maintain the elephant population at current levels through increased enforcement activity responded affirmatively 65% of the time. The average response was $89 while the median was $100.[10,11]

Some respondents dislike translating important qualitative experiences into a dollar metric and respond with a zero response. There were a substantial

[9] See 51 Federal Register, 27 674 (1986)

[10] **Special Fees and Permits**

(a) Suppose that the current population of elephants can be maintained if additional foot, vehicle and aerial patrols are operated on a sustained and regular basis in the parks. If these patrols can be supported by a special 100 dollar annual permit (or included in each visitor's safari costs), are you willing to support this permit fee?

[18] **NO,** I am not willing to pay $100 for this permit.
[34] **YES,** I am willing to pay $100 for this permit.
[ ] I am willing to pay a maximum of _____ for this permit. (Express this in U.S. dollars or specify in one of the currency equivalents in the table below).

[11] There is general agreement among specialists in contingent valuation procedures that a willingness to pay question is likely to generate a more accurate answer than a willingness to sell question.

number of zero responses which were not removed. Researchers can attempt to distinguish 'protesting' respondents from those whose zero is 'genuine' by introducing further questions in the survey. These questions were not added because it would have increased the length of the questionnaire and would have led, it was thought, to increased non-response because of the length of the survey.

Respondents can also behave strategically, giving zero values if they think the results will be used to design a policy change, or give large values if they think the results will lead to policy decisions which they like but won't have to pay for. Respondents may also put in large values if they regard the question as a sort of referendum in which they can vote, as it were, for a broader, perhaps moral issue. In this case, respondents may see themselves taking a stand on the preservation of species, not the smaller issue of maintaining elephants. The largest response to this question was $500, less than 1% of the respondent's annual income and about 3% of the cost of his safari. Some researchers remove observations if they exceed the respondent's income by more than 5 or 10%. According to this criterion for data trimming, no observation was removed. Unlike most other contingent valuation studies, no observations have been excluded. However, the importance of the zeros in this study has been diminished by using the median value ($100).

There is no compelling response to a cynic who argues: 'Ask a hypothetical question, get a hypothetical answer.' However, in a major test of a market value versus responses to a willingness-to-pay question for the same good, respondents understated their willingness to pay compared to the market value by a factor of about three (Bishop and Heberlein (1979)). It may be that the median value used in the present study similarly is understated.

On the other hand, respondents were not reminded of the other worthy causes they could contribute to, such as preserving the rhino. Moreover, there has been so much publicity about the crisis of the elephant population that the stated enthusiasm for paying might be tempered by the reality of actually writing out a cheque.

It may or may not be coincidental that the median value is identical to the value specified in the question. Survey researchers worry about a 'starting point' bias. Such a bias can be tested by varying the value in the question and testing the relationship of the response to the starting point. Inadequate sample size foreclosed this option.

The median value of $100 seems modest inasmuch as it is 3% of the total cost of a safari. If one thinks introspectively about the value over and above the cost of a very satisfying or extremely satisfying moderately expensive experience, $100 does not appear to be a suspiciously high number and some may think it is low.

Calculation of consumers' surplus

Combining the median value of willingness-to-pay of $100 with the estimate of 250 000 to 300 000 adult safaris per year yields an annual viewing value of elephants of between $25 and $30 million. If the mean value of $89 per person is accepted, the viewing value is decreased to between $22 and $27 million. These values are summarized in Table 9.3.

It is satisfying that the preferred travel cost approach (with positive time value) produces values ($23 to $27 million) very close to the contingent valuation methods values based on the preferred median value, or $25 to $30 million. However, confidence about the convergent values should be tempered by the sample size of 2 experiments. Nevertheless, these estimates are the best there are at the moment and will have to stand until more research in this area is undertaken. It is also important not to belittle these estimates too much. They are almost certainly a good guide to the order of magnitude of value. For all the fragility of the research, it is fair to conclude that the viewing value of elephants is more like $25 million annually, than $2.5 or $250 million. The figure of $25 million in Table 9.3 is selected as a best point estimate, merely because it falls within the range of the preferred estimates. In view of this estimate, Kenya's 1988 Wildlife Management and Conservation budget of under $200 000 seems not to be excessive.[12]

9.4 SAFARI EXPENDITURES IN KENYA

It is popular among advocates and typically quite misleading to represent expenditures on a particular good or service as the value society would lose if that good or service was removed. It was pointed out earlier that consumers would react to the removal of a good by spending their money elsewhere. However, from the perspective of Kenya or any other African country, the loss of tourist revenues due to declining elephant populations is very likely a loss of employment and profit to that country. Only order of magnitude values are possible to make. The US Embassy (unpublished and undated) estimated that tourism and coffee generated about $280 million each in 1988. Not all tourists go on safari. If for consistency with US Embassy data, 60% of the tourists go on safari, then safaris earn about $168 million in foreign exchange.

Consider an alternative way to estimate expenditures. Our surveys indicate that the average land cost for a safari is $1033, using a weighted average for North America and Europe; and it was assumed that 250 000–300 000 tourists went on safari. These figures result in expenditures of between $258 and $310 million annually. An unknown fraction of

---

[12] Estimate obtained from an unpublished memorandum from US Embassy.

Table 9.4 Domestic loss of tourist expenditures if elephant populations decline

Further loss of elephant populations (%)	Loss of expenditures (millions of US dollars)
25	52–63
50	86–103

land costs is spent outside of Africa. Profits for large tour operators, financially based outside of Africa, certain administrative and advertising costs and purchases of durable equipment such as transportation vehicles are examples of important leakages from the country. Our guess is that the leakage fraction is not much less than 20% or more than 50%. An estimate of 35% produces domestic land cost expenditures of between $168 and $202 million. This range is accepted since it includes values obtained by two different approximation procedures.

What is the likely consequence of poaching activity on safari expenditures, either because of the direct impact on elephant populations or out of personal fear of poachers? Continuing the adventure into unchartered estimates, results of the tourist survey indicate that 31% and 51% respectively would not recommend safaris to their friends and relatives under circumstances of a 25 or 50% decline respectively in elephant populations. Applying these percentages to the earlier estimates of domestic expenditures indicates a loss of over $50 million to Kenya for a 25% decline in elephant populations and a loss of more than $85 million to Kenya if the elephant populations decline as much as 50%. These estimates are summarized in Table 9.4.

REFERENCES

Bishop, R.C. and Heberlein, T.A. (1979) Measuring values of extra market goods: are indirect measures biased? *American Journal of Agricultural Economics*, **61** (5), 926–30.

Clawson, M. and Knetch, J. (1966) *Economics of Outdoor Recreation*, Johns Hopkins University Press for Resources for the Future, Baltimore.

Cummings, R.G., Brookshire, D.S. and Schulze, W.D. (eds.) (1986) *Valuing Environmental Goods: A State of the Arts Assessment of the Contingent Method*, Roman and Allanheld, Totowa, NJ.

Federal Register (United States), 51, 27,674 (1986).

Kenya for Republic of Kenya, *Economic Survey* 1989.

Kenya for Republic of Kenya, *Statistical Abstract*, 1986.

Mitchell, R.C. and Carson, R.T. (1989) *Using Surveys to Value Public Goods: The Contingent Valuation Method*, Resources for the Future, Washington, DC.

U.S. Water Resources Council, (1983) Economic and environmental principles and guidelines for water and related land resource implementation studies, US Government Printing Office, Washington DC.

10

Ecology and economics in small islands: constructing a framework for sustainable development

Stephen M.J. Bass

10.1 INTRODUCTION

The potential for small islands to pursue sustainable development depends upon maintaining the quality of their necessarily limited natural resources. At their most basic, these resources provide essential life-support systems: maintaining water supplies and soil fertility, and protecting the island from coastal erosion. Yet, historically, many small islands have developed by liquidating natural capital, a process having its origin in the 'frontier' culture of western economies. As a result, the natural life-support systems – for which there are few substitutes – are now critically diminished.

Islands may be viewed as comprising mutually inter-dependent subsystems: economic, social/demographic, cultural, political, physical and ecological. The interaction of these subsystems defines the behaviour and sustainability of the island in the face of external influences and internal adjustments. A sustainable equilibrium is achieved when each subsystem performs acceptably, resulting in increases in income, health, cultural richness, island decision-making autonomy, biological diversity, and – as we have noted – secure ecological life-support. Disequilibrium results when stresses are so high that one type of society, economy or ecology replaces another too rapidly, with inadequate time for all the subsystems

Economics and Ecology: New frontiers and sustainable development.
Edited by Edward B. Barbier. Published in 1993 by Chapman & Hall, 2–6 Boundary Row, London SE1 8HN. ISBN 0 412 48180 4.

to adjust. Such disequilibrium has plagued the history of many islands (McElroy and de Albuquerque, 1991).

That disequilibrium is prevalent is largely due to the high exposure of island ecologies, economies and societies to external influences, and the low capacities for adjustment in relatively small, resource-poor islands. Disequilibrium is especially pronounced in the smallest islands: those with less than 800 km^2 and fewer than 100 000 people (Brookfield, 1986).

Small islands may be characterized by:

1. economic dependence on larger countries for markets and investment and, most significantly, for sea and air transport;
2. inability to exploit land transport fully;
3. geographic isolation (which, however, can effectively be reduced by proximity to an established sea or air route);
4. small populations, and hence a limited pool of skills;
5. yet often high population densities, and hence high demands on resources (Hong Kong, Singapore, Malta and Barbados have some of the highest population densities in the world);
6. highly circumscribed space; paucity of natural resources, and hence low resilience;
7. the intimate linkage of all island ecosystems: impacts in one part will affect other parts;
8. high ratio of coastline to land area, leaving islands vulnerable to marine and climate influences, such as cyclones, hurricanes, storm waves, salt-related corrosion and marine pollution;
9. vulnerability of island ecosystems to other external ecological influences, notably exotic species introduction.

These characteristics are fundamental parameters for small island development; yet development has tended to proceed with inadequate information on them.

Disequilibrium in small islands has been most clearly exhibited in the string of commodity booms and collapses which have characterized island development thus far, e.g. in minerals, timber, sugar, bananas, migrant labour and more recently tourism.

Island economic growth has 'taken off' through exporting natural resources of highest value at the time. These may be minerals, e.g. phosphate in Kiribati (now exhausted) and Nauru from the mid-1800s, nickel in New Caledonia from 1870 and oil in Trinidad from 1900. High value timber stocks were devastated in much of the Caribbean, e.g. Honduras mahogany from Barbados from the 16th century, parts of the Pacific, e.g. sandalwood from Hawaii from the late 18th century, and the Indian Ocean, e.g. ebony from Mauritius. These wholesale removals of very particular natural resources have been accompanied by substantial environmental degradation – due in part to the high interconnectedness and fragility of island ecological systems, and significant social changes – due in part to the small size of island social systems.

Mineral extraction, timber removal, and the establishment of plantation crops, have all led to the most significant anthropogenic environmental transformation of islands: deforestation. The typical process has been:

1. logging of valuable hardwoods;
2. replacement of natural forests by erosive and pest/disease-prone plantation monoculture;
3. soil exhaustion and plantation epidemics;
4. plantation collapse and the subsequent marginalization of poor people to upland forests;
5. upland deforestation by displaced people;
6. consequent upland erosion, and hence further appropriation of the island's natural ecosystems – and subsequent further degradation;
7. economic disinvestment of the degraded island interior in favour of coastal development or, where environmental degradation leads to collapse of life-support systems, emigration;
8. extreme lack of investment in managing the ever-diminishing forests.

The process of deforestation has been exacerbated by natural disasters such as hurricanes which periodically destroy the vulnerable plantation monocultures and remaining forest; by the neglect of 'unprofitable' islands on the part of colonial and dominant economic powers; and by the inability of small island populations to muster the skills and the political and economic power to counteract the deforesting trend. As a result, biological productivity, diversity and resilience have diminished. Much land in many islands now lies derelict.

Caribbean development is historically associated with tremendous forest loss, a loss which now has disturbing ramifications for the future of island economies and ecologies. Large stands of Honduras mahogany were logged from the 16th century onwards, and were logged out by the early 19th century. From 1630 to 1880, larger areas still were deforested for export plantations, notably for sugar and cotton. As early as 1655 in Barbados, only very small forest relicts remained, and by 1700, the soils of some estates were severely impoverished in attempts to maximize the production of sugar (at that time considered to be a valuable spice) (Watts, 1987). Today, Barbados has less than 1% forest cover, one of the lowest in the world.

The inevitable consequence of land degradation was abandonment. It was hastened by the plummeting economic viability of plantation estates in the mid- to late-19th century, due to: rising labour costs after the emancipation of slaves; price competition from larger countries producing sugar and cotton at lower cost with economies of scale not available to islands; and, in the case of cocoa and citrus estates, disease epidemics which quickly covered the small islands. Other land was abandoned due to the massive emigration that has occurred over the last 160 years, and especially to the United States and the United Kingdom since the 1950s.

Plantation collapse and slave emancipation increased smallholder pressure on land not held by plantation owners, notably upland forests (Watts, 1987). Settlement on steep, forested slopes continues today, partly because smallholders – most of them still legally termed squatters – are debarred from flat land held by major landowners.

In summary, Brown (1982) has shown that the forest area in a given Caribbean island reduces with increasing population, higher energy consumption (as a measure of human activity), increasing road networks and lower, flatter topography. Today, half the world's nations with forest cover below 5% are small islands. Yet, despite such low forest cover, the values of remaining intact forest for supporting other island systems – water supplies for agriculture and domestic use, landscape for tourism, etc. – are little known or appreciated.

Some islands this century have responded to this legacy of degradation by developing an extreme form of economy independent of local resources but dependent upon economic opportunities elsewhere: the 'MIRAB' (migration, remittances, aid and bureaucracy) economy. Migration may be massive in response to employment booms abroad: there are more Cook Islanders, Tokelauans and Niueans living in New Zealand than there are in these same islands (Hamnett, 1986).

Other islands, especially those with very few natural resources, have adopted a development path that depends upon income from banking, insurance, postage stamp sales and tax-free financial markets – income which accrues simply by virtue of the island's identity. Together with subsidized, imported food and an ageing workforce, they explain the paradox of land remaining abandoned and unrestored in so many overpopulated small islands. Both these forms of development avoid many of the risks of environmental degradation experienced with the (earlier) exploitation of forests and minerals and agriculture. Although some believe that they could possibly represent durable development models (e.g. Ogden, 1989), such development models remain subject to disequilibrium, in part because environmental management is neglected, and hence essential life-support systems are vulnerable.

MIRAB economies are only one form of the extreme social transformations that have resulted from externally-induced changes in islands. Most extinctions of human populations (as well as the often-cited animal and plant species) have been on islands: the Guanches in the Canaries, and the Arawaks and the Caribs in the Caribbean (Crosby, 1986). Only where there has been insignificant colonial influence have indigenous cultures survived, such as many of the Pacific Islands, the Maldives and the Comoros (Hein, 1986). Throughout the Caribbean, however, today's population is the product of colonial settlers and administrators, African slaves, and indentured East Indians and other labourers brought in to take the slaves' place following emancipation. No Arawaks and almost no Caribs survive. The new Caribbean people have adopted cultural and demographic means to create as much resilience as possible, e.g. the races mixing freely to establish a national culture; emigration and immigration according to economic and environmental fortune. Yet they have developed few resource management practices that are sustainable, in contrast to other islands where 'pre-contact' populations and resource practices remain today.

Today, sustainable development continues to elude most islands. Structural adjustment has created overwhelming imperatives to export. Many islands are attempting a transition from boom–bust exports dependent upon agriculture and natural resource exploitation to tourism and industrial exports. The new activities also depend upon environmental resources, however: watersheds, landscapes and coastlines; and they also have environmental impacts: solid waste, pollution, landscape change, erosion of cultural traditions and excessive appropriation of natural habitat. Like deforestation, such impacts can be irreversible and limit the performance and resilience of island subsystems.

These new island problems often have roots which are similar to those of deforestation: that of a 'resource frontier' approach to development. This approach is ingrained in island policy, administration and enterprise, following the precedents of (colonial) continental experience. The behaviour of island ecologies and island economies, and especially their interactions, is little explored. Such an exploration, however, will be fundamental to defining sustainable development in islands. What must become clear is that, in islands, there are few expanding 'frontiers': there are limits to resources and to sinks for wastes.

In this chapter, we argue the need for approaches to island development that are informed both by insight into the peculiarities of island economies and island ecologies, and by the traditional methods of operating within island constraints. We focus on small islands in tropical regions. We outline the requirement for a strategic planning and monitoring framework for sustainable development in small islands. Such a framework might encompass:

1. systems for analysing the island's economic and ecological characteristics, and especially their limits;
2. systems for analysing external economic and ecological influences on the island;
3. systems for assessing the interactions of the island's characteristics with the external influences;
4. involving public participation in decision-making and resource control measures, to maximize positive interactions and minimize negative ones;

5. instituting traditional and new resource management systems for restoring, stabilizing and developing the resource base;
6. establishing hazard management capabilities;
7. generating multiple use possibilities for island resources;
8. equipping institutions and staff to undertake multiple functions;
9. reducing the isolation of island professionals;
10. establishing policy and economic incentives and other means to achieve the above.

10.2 THE VULNERABILITIES OF ISLAND ECONOMIES AND ISLAND ECOLOGIES

Today, in all countries of the world, there is a growing understanding of the linkages between economics and the state of the environment. Furthermore, it is widely accepted that this understanding must inform the political and development processes. For small islands, therefore, knowledge of the peculiar characteristics of island economies and island ecologies, and their interactions, is vital for sustainable development.

10.2.1 Island economies are based on a narrow set of activities, and are highly exposed to external economic influences

Small islands seek autonomy, yet they can develop only through interacting with larger economies, which provide capital, markets, and transport links. However, external trading partners are rarely 'captive' to an individual island, and hence have not been induced to invest in the long-term diversification and security of island enterprises.

Small islands have limited capacities both to produce and to consume; they cannot create monopolies and operate large-scale operations; they cannot develop substantial internal markets; and they cannot raise large amounts of capital/finance on the home market. The economies of scale necessary to cover high transport costs are elusive; and islands become dependent upon transport links with just one or two countries – and hence on a limited market. Few islands ever record a trade surplus. They import inflation, and exchange rates are beyond their control, frequently being pegged to the dominant trading partner's currency. Economic restructuring is hampered by the excessive mobility of crucial domestic determinants of future growth – capital and labour (McElroy *et al.*, 1987). Where labour has been used to migrating, it comes to expect high wages, which cannot always be sustained.

A feature of island economies is the prevalence of boom and bust, based on very few commodities, with the boom phase being inadequate to permit any diversification (*The Economist*, 1988). Islands need to diversify, and yet there is inadequate manpower and resources available to do so. And when they manage some diversification, very often the most that can be achieved is to produce small quantities – merely 'samples', as one British aid official has described it to this author.

With non-renewable resources, the problem is the misfit between the optimum rate of depletion to an external investor (who can move on to another source) and the optimum for the host society, including its future generations. Yet the need for foreign exchange, and the fear that technology may remove their markets in time, has meant that islands have permitted external investors to keep the upper hand, and hence to call the shots in depleting resources (Girvan, 1991).

Small islands are price-takers and hence also 'become chronic takers of technology, infrastructure, resource-allocating institutions, trade patterns – in brief, all of the strategic decisions that circumscribe their viability are externalised' (McElroy *et al.*, 1986). Many small islands also receive a high proportion of aid or transfers (Hein, 1986). In 1984, aid as a percentage of GDP was 43% in Kiribati, 84% in Niue and 118% in Tuvalu (Ogden, 1989). In short, there is a frequent discrepancy between the legal status of independence of many islands and their practical needs for dependence (Knight and Palmer, 1989); and opportunities to evolve legitimate island approaches to development have not arisen.

The exception to the general rule of economic dependence arises if countries can raise enough capital to seize locational advantages, e.g. as Singapore did in the 1960s. However, scale and locational constraints (or occasionally, as with Singapore, advantages) are not permanent parameters but may change with new technology. Mauritius was the 'star and key of the Indian Ocean' in the 18th century, but its locational advantage between Europe and Asia was removed once the Suez Canal was opened (Brookfield, 1986).

In practice, most islands become linked with (often very distant) poles of political and economic importance and, even within a small archipelago, individual islands may be associated with different foreign influences. Often the former colonial power – which was responsible for the island entering the international economy in the first place – is the dominant influence, e.g. in the Caribbean, the predominant trading partners include the UK, France, The Netherlands and the USA. Hence it is common that regional economic interests do not coincide unless islands in a region share similar relationships with a major country. As a result, regional trade can be relatively slight. For example, in the early 19th century there was considerable hope that Trinidad, experiencing decline in its sugar industry, would be able to take advantage of its proximity to South America and other islands – effectively to become a 'Liverpool in the West Indies'. The political, economic and transport links with South America were too weak, however, and the fact that the new South American republics were so unstable meant that the interests of Great Britain remained predominant (Wood, 1968).

Furthermore, the frequently great distance of small islands from their dominant foreign partners, and the ability of the foreign power to choose alternative partners if offered better deals, mean that islands suffer a lack of stability in investment, aid and markets. For example, the Caribbean sugar and tourism industries – the two most significant in that region – are driven by the North American economy and by its political policies towards the Caribbean. When there has been recession in North America, or when political events in the Caribbean have not suited the USA, island economies have suffered.

Changes in the internal policies of trading partners can also upset island economies. For example, the favourable market which Windward Islands bananas currently enjoy in the UK is threatened by the prospect of a single European Community market at the end of 1992, when the Windwards may have to compete with far cheaper bananas from elsewhere.

Since World War II, tourism has exerted the most profound influences on tropical islands (Doumenge, 1985). The smaller islands of the Caribbean (below 800 km^2) produce only 7% of the Caribbean's GDP, but now account for 40% of tourism expenditure (McElroy and de Albuquerque, 1991). Many islands possess a comparative advantage in tourism, because of their high proportion of coastal land and unclouded skies. Although tourism does not directly consume island resources, it is highly dependent upon markets and political conditions in market countries. Furthermore, the environmental conditions upon which tourism depends are themselves influenced by foreign markets; these encourage deliberate changes in the environment , e.g. through built development, as well as leading to many indirect changes.

Indeed, McElroy and de Albuquerque argue that the 'overgrowth propensities endemic to island policy and international tourism interests' seem inevitably to lead to major ecological and social transformation. Tourism development in virtually all small Caribbean islands appears to have made an inexorable progression towards high-density, mass-market styles which entail intensive infrastructure development on the vulnerable coast, high attendant social, cultural and ecological impacts, and almost total macroeconomic dependence upon external tourism markets (McElroy and de Albuquerque, 1991). In the face of dominant foreign influences and the need to generate foreign exchange rapidly, island tourism industries are unable to self-impose an optimum scale of operation consistent with social and ecological sustainability (Daly, 1984, quoted in McElroy and de Albuquerque, 1991).

Clearly, if its scale is not controlled, tourism presents a potential environmental hazard, especially as it is concentrated on the relatively fragile coastal ecosystems. It can be socially destabilizing, for in many islands the tourist population greatly outnumbers locals. Sustainable tourism will depend upon integrated policy and planning that will maximize visitor expenditure rather than numbers, that will widen the season but also allow for ecological and infrastructure recovery, and that will provide measures to control ecological and social impacts.

Finally, it must be noted that, for many small islands, the surrounding marine territory may be the most important economic resource. This territory – which is often far larger than the island – must always be considered as part of the small island system. Nearshore fisheries provide subsistence, social security reserves and unemployment insurance for many island people (Office of Technology Assessment, 1987). However, they depend upon very limited areas of seagrass beds, mangroves and reefs, which are easily degraded by industrial and urban development – particularly pollution and land reclamation. Offshore fisheries present far larger resources for development, but they often require technology and capital from other countries; again, therefore, island economic resources are placed in the hands of external powers.

10.2.2 All the ecosystems on a small island are intimately connected

Small islands are characterized both by a high diversity of terrestrial and marine ecosystems per unit-area, and by the extensive links between these ecosystems. These links provide valuable mutual support between ecosystems, and they contribute to life support mechanisms. For example, upland forests buffer lowland systems by slowing rainwater runoff and erosion, and by recharging streams and groundwater supplies at a rate which is steadier than that of rainfall. By moderating the movement of sediments, forests also renew the biological productivity of savanna croplands, marshes and mangroves. These low-lying systems further trap water and sediment to protect the clarity, and buffer the salinity, of marine lagoons and coral reefs. Hence good management of one agroecosystem, such as upland forests, can improve the management of other agroecosystems, such as agriculture, tourism and pelagic fisheries – the interaction of these activities with forestry being greater in islands than in other systems. There are also certain natural energy 'buffers' to reduce the inherent ecological vulnerability of islands to external influences. For example, coral reefs and mangroves which fringe islands protect their shores from storm surges.

Yet, by virtue of these same links, an ecological event in one part of an island can have consequences in another (Office of Technology Assessment, 1987). The Caroni Swamp in Trinidad provides many examples of ecosystemic interlinkages – and their economic consequences. Until very recently, this mangrove swamp was treated literally as a wasteland, and much of Port-of-Spain's solid waste is still dumped there. This waste, as well as industrial effluents and agricultural chemical run-off from sugar estates, have resulted in high toxicity levels damaging to mangrove fish nurseries. This problem is exacerbated by nearby port and road construction and by deforestation in the mountainous Northern Range, which increase the swamp's silt load. The biological diversity of the swamp is decreasing, which eventually will damage its most financially rewarding direct use – ecotourism. Eventually, if the swamp continues to diminish, inland systems will be at greater risk from storm surges and possible sea level rise.

10.2.3 Island ecologies are highly vulnerable to external environmental influences

The ecological significance of small islands is disproportionate for their size. This significance arises from:

1.   the small populations of island species;
2.   their isolation from alternative populations of the same species;
3.   the limited array of competitors and specific predator/prey relationships.

Many islands are important areas of biological endemism, and island ecosystems commonly exhibit high species richness. Yet some of the factors which render islands of ecological significance also leave them susceptible to external influences. Islands are often incorrectly exemplified as 'closed systems' when in fact they are often very far from this; and they are becoming increasingly open every day. Small islands are particularly vulnerable to coastal and marine influences.

The survival needs of the densely-packed human populations (which often depend upon imported biological diversity such as cattle, goats and annual crops, and engage in environmentally-damaging occupations) have for long overridden longer-term concerns such as the conservation of indigenous biological diversity. It is not therefore surprising that most of the recorded species losses of the last few centuries have been in tropical islands. Many species became extinct shortly after colonial conquest, with the introduction of colonial crops, livestock and their attendant panoply of pests and diseases (Crosby, 1986). Many of the island species that remain have lost their competitive ability, and whole island ecosystems are vulnerable to collapse if exceptionally invasive species are introduced. There are one hundred times more endangered species *per caput* in the island Pacific than in mainland Africa, for example (Dahl, 1984). Island resource management has to be exceptionally carefully designed if it is to sustain island biodiversity; yet there are few special provisions or guidelines available for this.

Islands in the Caribbean and the South Pacific are especially prone to natural disasters such as hurricanes, cyclones and volcanic eruptions. The smallest islands cannot deflect hurricanes and cyclones, and they are not large enough to moderate general climate circulation patterns. This renders them vulnerable to drought and other climatic events, which can destabilize complete ecosystems. Certain island ecosystems are resilient to such events e.g. 'hurricane forests', which regenerate following frequent hurricanes. However, where island ecosystems are not left wild but are used for human purposes, such long-term ecological resilience is inadequate, and natural disasters can be highly damaging for human enterprise in the short term. They erode the productive resource base, and natural regeneration is not speedy enough to restore essential ecosystem processes.

Hurricanes are the most significant natural calamities in the island Caribbean. In 1955, Grenada's then-biggest industry, nutmeg cultivation, collapsed. The banana plantations in Saint Lucia were destroyed in 1980, as were five million forest trees in Dominica. The direct costs of Hurricane Gilbert in Jamaica in 1988, which devastated the fruit, coconut and coffee industries, amounted to US$956 million, according to the Planning Institute of Jamaica. Half of the losses were in agriculture, tourism and industry, and half in housing and infrastructure. Instead of a growth in GDP of 5%, a decline of 2% was now expected (Vermeiren, 1991). What is not clear is whether the Planning Institute's cost estimation included damage to the natural resource base in addition to man-made resources – from e.g. landslides and deforestation in watershed catchments, and their effects on soil and water quality and quantity. If the Institute had followed the procedures of the UN Disaster Relief Organization in cost assessment, which is likely, then these costs would not have been included (Vermeiren, 1991). (The persistence of sugar over the centuries – although it is now a low value crop with increasing input costs – is in part due to its relatively high resistance to hurricanes.)

In future, it is possible that the natural energy buffers, which protect islands from damage from the sea, may be tested beyond their limits. The sea level rise and increased storm frequency expected with probable global warming may inundate much of low-lying islands such as Anguilla, Kiribati, Vanuatu, Tuvalu and the Maldives. In the Eastern Caribbean, for example, there may be a rise in average temperature of 1.5°C by 2030 and slightly increased rainfall and humidity. Such a temperature increase at the surface of the sea could increase the frequency of hurricanes by 40% and the maximum wind speed by 8%. Furthermore, a 30 cm sea level rise is expected by 2030. This will all lead to an increase in wave energy and destructive power, with the following physical effects (Bass and Cambers, 1991):

1.   inundation of coastal low terrain;
2.   increased beach and cliff erosion;
3.   migration and/or reduction of wetlands;
4.   saltwater penetration;
5.   altered tidal currents.

The net result of all these impacts is likely to be disequilibrium in social and economic systems. In the Eastern Caribbean, most of the economic resources (particularly for tourism) and the strategic resources (infrastructure such as ports, airports, roads, housing and services) are coastal. They are highly vulnerable to rising sea levels and storm surges. This will bring with it higher costs of insurance and/or failed development. It may become increasingly difficult to attract and insure investment, particularly foreign investment and especially for coastal developments. The comparative advantage of areas which are not sensitive to climate change, and particularly to sea level rise, will increase. This will affect land prices and alter land use patterns, reversing the current trend towards investment and settlement in the coastal zones and shifting population groups to more fragile uplands. As yet, however, there are inadequate planning frameworks to ensure that new investment will be sustainable (Bass and Cambers, 1991).

10.2.4 Deterioration of island environments erodes indigenous economic and social potentials

Natural ecosystems, particularly forests, are exceptionally important for island societies, even though their total area may be minuscule on a world scale. Small island forests can be absolutely critical for life support systems, especially for local water supplies, subsistence wood supplies and soil stabilization. The watersheds of small islands are far smaller than those on the continent; even slight forest degradation can destroy watershed functions. In some small islands , e.g. Carriacou, the forest has been cut to such an extent that fresh water has to be brought in by ship (OAS, 1988) and in others, e.g. Curacao, energy-intensive desalinization plants are needed. In one island off Bougainville, almost all wood and water has to be shipped in from the mainland. The loss of forests in Trinidad has led to widespread flooding of homes and farmland in the wet season, and erratic water supplies in the dry season; the costs of this forest loss are estimated at hundreds of millions of dollars per year in an island with a population of 1.3 million. Likewise, island aquifers are small and vulnerable; even minor saltwater or toxin intrusions can render an aquifer unfit for human use – at great cost to local people and the tourism industry. Even where aquifers are well-managed, they may simply not yield enough to meet demands. In Kiribati and the Marshall Islands, for example, atoll aquifers are already inadequate to meet the needs of growing urban populations (Hamnett, 1986).

10.2.5 Small islands suffer many constraints in tackling resource degradation and unsustainable development

Firstly, there is currently little incentive to address unsustainable resource use. The total values of resources such as beaches, coral reefs, wetlands, aquifers and forests are not signalled in economic terms, as we have seen above. Many are treated as free resources locally, and the incentive is to liquidate those which are of the highest value internationally.

Second, island manpower and skills are inadequate to tackle the problems of unsustainability. The public sector plays a disproportionate role in the economies of island nation-states. In the Pacific, government expenditures range from 30 to 50% of GDP (Hamnett, 1986). In the Eastern Caribbean, governments employ 30% of the labour force. A priority for nearly all governments is to reduce the proportion of the labour force which does not generate income. Combined with the incentive to liquidate resources, this has meant that seemingly long-term concerns such as environmental management have been neglected. Small islands cannot afford to train and maintain the numbers of forestry, agriculture, water and conservation professionals that would be required to manage island resources under 'continental' sectoral models of resource management. Consequently, many island governments will employ only one of each professional, if at all, in a vertical hierarchy which inadequately recognizes the links between island ecosystems. A more integrated approach to natural resource management is required.

Most island professionals will, by necessity, have been trained outside the region. Yet approaches developed for large, continental areas are not always appropriate. There are very few natural resource management training institutions in island regions, and training does not address the wide range of disciplines required for resource management in small islands. For example, a small island forester is potentially involved in watershed management, tourism, agroforestry, wildlife conservation, timber import and export, and public participation, as well as timber production. Training in the latter alone is no longer appropriate.

Third, immediately-available funding to tackle unsustainability has been low. It is not possible to raise funding from external trading partners that are not 'captive' to the island, nor are local resources adequate. Significant aid funding has noticeably been focused on the larger continental countries. The special case of island resources – small in quantity but great in human significance – has gone unnoticed.

Fourth, inter-island co-operation, potentially of value in dealing with the limitations of micro-states, is frequently weak, because of poor inter-island transport, trade and joint facilities, language problems, etc.

Fifth, as we shall explore in the next section, even in islands without a colonial heritage, external economic and social influences are now so high that potentially restorative traditional resource management systems are disappearing. These have been pushed aside by people who are no longer satisfied with 'subsistence affluence' but who have higher material aspirations; they have been discredited by similarly western-looking authorities; or they are no longer effective with higher population and economic pressures. Societies in most islands have readily accepted what Girvan

(1991), in observing the Caribbean, complains has been 'the tyranny of the ideology of GNP growth associated with capitalism and modernisation theory'. Progress has been equated with the 'indefinite accumulation of material artifacts, and the environment is assumed to be infinitely capable of supporting this' (Girvan, 1991).

Finally, it should be said that there are also certain advantages to being small, notably the potentials to reach consensus, to monitor environment and development more closely, and to adapt more speedily. Also certain problems that face larger countries, such as large-scale industrial pollution, are much reduced.

10.3 ISLANDS AS OPPORTUNITIES FOR SUSTAINABLE DEVELOPMENT

If island development is to become sustainable, the fundamental reason for historically unsustainable development must be explored. This is the externally-imposed – but now also internalized – 'resource frontier' approach to islands.

Before the colonial era, Europe's agriculture and industry depended upon intensive techniques, because of its increasingly restricted resource base. The 'new world' of the colonies, in contrast, opened up rich possibilities for expansion. In ensuing centuries, predatory natural resource use came to characterize imperial European civilization: a process of land territorialization, capitalization and abandonment, and transfer of economic activity to further 'virgin' tracts. The success of this approach led some emergent settler cultures to cultivate a mythology of 'expanding frontiers'. The profits helped to finance urban-based industries; and the dynamic of industrial growth served in turn to sustain the mythology of unlimited frontiers, and further transformed formative frontier myths into a belief in perpetual economic growth.

> Having expanded on the things of nature, the West came to believe
> that expansion was in the nature of things. (Weiskel, 1990)

This frontier approach is now prevalent throughout the world, in development policy and in the conduct of business. Even small islands – in the grip of larger countries and corporations – operate in the same way. Islands were easier to colonize and control than continental countries, and much of the earliest colonial frontier agriculture was based on islands. Yet islands command inadequate space to sustain such an approach, and environmental degradation and social transformation has been evident from soon after the earliest colonial conquests (Wood, 1968 and Watts, 1987).

Some circumstances, however, have enabled some islands still to uphold frontier myths. Recently, 200-mile exclusive economic zones (EEZs) have been declared around many islands. EEZs have pushed back the resource 'frontier' considerably. For example, the Pacific island tuna resource that now lies within this zone is the largest in the world; and the catch, insignificant before 1970, increased to 35% of the world catch in 1984 (Hamnett, 1986). Similar expansions of 'frontiers' – and of short-term fortunes – have been experienced in the Falkland Islands.

Other attempts to push back the frontiers have been less successful. For example, urban infrastructure in the Maldive Islands has been constructed on new land made with coral dredged from the sea. The costs of exploiting this 'free' resource may yet prove to be far higher than those of its extraction. The coral reef is an important buffer for storm waves, of increasing importance in the face of projected sea level rise and greater storm surges, and its removal may already have contributed to increased incidences of inundation (Ince, 1990).

In general, therefore, it is not possible for islands to create new resource frontiers. The challenge for islands is to create policy, institutional and technical frameworks for 'post-frontier' development: for sustaining the natural resource capital, living off the 'interest' produced by this capital, and closing ecological cycles so that wastes become resources, and resources are renewed. Initially, to do so will entail uniting efforts of economic development and environmental management. Ultimately it will depend upon changing the premises and beliefs of island societies. To do both will require greater insight into the peculiarities of island economies and island ecologies, and their interaction with external influences.

Insight might also be sought from traditional responses to island circumstances: the resource management techniques and control systems associated with pre-colonial societies. For example, prior to European contact, Melanesian and Polynesian island societies exploited natural resources mainly to meet subsistence needs. On fertile, volcanic islands a 'subsistence affluence' was achieved; its food and labour surpluses were employed to support elaborate religious and political systems, not to develop resource-exploitative export-driven economies as they were on the colonized islands (Hamnett, 1986). Even on atolls with very poor soils, sustainable subsistence systems were established. In Polynesian atolls, the concept of *raui* (restraint) prohibits resource exploitation by closing off areas or seasons, limiting exploitable sizes or harvest levels. Such control systems, and resource ownership, are continually negotiable through traditional decision-making structures. They are hence far more resilient, and take a more 'holistic' view, than their western equivalents. And, just as significantly, many of these control systems are based on the premise that resources are held in trust for future generations. There are traditional forms of land management, too, that can ensure sustainable yields even on infertile atolls, e.g. mulched taro pits, dug below the water table; and there are traditional land capability classification systems that recognize island ecological dynamics (Liew, 1986).

Such insights into islands may possibly reveal lessons for the world as a whole. To an extent, islands provide microcosms of global issues – for these are clearly places where it is impossible to escape indefinitely the reality of resource limits. Indeed, the germ of a colonial realization of the possible limits to resource frontiers evolved in tropical islands such as Mauritius, St Helena and the West Indies, as early as the late 17th century. Grove has noted that the work of the medical practitioners and botanic garden curators posted to these island colonies was, in fact, fundamental in forming later conservationist responses in Europe. Tropical islands – once considered as earthly paradises, later explored and catalogued scientifically, and finally exploited for their natural resources, with consequent environmental crises – became allegories of the whole world. Men of vision were able to deduce from colonial island development experiences what might occur at the global level if development patterns proceeded as if resources were limitless. Many of their arguments were as mature as those posed by environmentalists today (Grove, 1990).

10.4 FRAMEWORKS FOR SUSTAINABLE DEVELOPMENT

The economic demands placed on island resources, particularly by dominant economic partners, are frequently overwhelming and destabilizing. The susceptibility of small islands to the environmental problems associated with these demands, and their vulnerability to other external climatic and environmental influences, are both high. In contrast, the capabilities of most islands to analyse these issues and to plan appropriately are limited. Hence the reactive approach of many islands, whose major strategic decisions are effectively made externally.

To tackle these problems, there is a need to:

1. understand the island's economic and ecological characteristics: the resource capabilities and their economic and ecological carrying capacities (by landscape/seascape units); and the economic values of resources and life-support processes;
2. understand the rationale and operation of traditional resource management systems, and the preconditions for their successful use;
3. understand the external economic and ecological influences: their type, degree, frequency/hazard, and likely costs/benefits;
4. assess the island's interactions with external influences: changes to capability, carrying capacity and values; and identify ways of maximizing positive interactions and minimizing hazardous ones so as to maximize resilience;
5. understand current resource uses, and the economic and policy signals encouraging those uses;
6. set strong strategic 'bargaining' positions with external decision-makers, to counter the boom–bust syndrome;

7. institute public participation in decision-making and resource management, based on traditional systems where appropriate. Most important (as Girvan, 1991, points out) is achieving national consensus on priorities where there are few resources and little time and technology to deal with an overly-comprehensive agenda; and agreeing cooperative state/private/community responses;

8. set ultimate limits to the appropriation of natural habitat, based on the total economic value of natural habitat – direct and indirect use values, and non-use (option and existence) values, the costs of its removal, and the costs of its conservation. The provision of essential life-support systems, such as watershed conservation and coastal protection provided by natural habitats, is likely to figure prominently;

9. restore, stabilize and develop the resource base outside protected natural habitat – to include strategies aimed at multiple and adaptable purposes, using minimal external inputs, closing ecological cycles, combating wastage and taking advantage of renewable sources of energy ;

10. establish hazard management capabilities, e.g. for oil spills, hurricanes and other potentially catastrophic external influences;

11. equip institutions and staff to undertake multiple functions, and to have a good grasp of the interactions of economic, ecological and social subsystems, especially in the smallest islands;

12. enable training institutions to produce highly-qualified generalists;

13. establish policy, legislative, participatory and economic incentives and other means to achieve the above.

A strategic approach can combine the above requirements in reorienting island development towards sustainability. It would have four components. One, a framework for planning and public participation to forge new directions; two, tools for analysing the island circumstance and generating solutions within such a framework; three, resource management techniques appropriate for islands; and four, institutional strengthening. We discuss the first three below. Institutional strengthening would follow forms derived from the specific interactions of these three.

10.4.1 Developing the strategic planning framework

An island's strategic planning framework needs to be participatory, for fundamental decisions about the future of island societies will be made. For example, should development policy aim at nurturing self-reliant nation-states irrespective of the personal sacrifice this may entail for residents, or should it aim at material development, perhaps at the cost of some island autonomy? (Ogden, 1989). The planning framework also needs to encompass far more concerns than the traditional frameworks of physical or economic planning, because of the extensive interactions of island subsystems. There are few precedents. National conservation strategies (NCSs) and similar initiatives would appear to be the closest equivalents. For example, in national conservation strategies, parallel processes of professional and public consultation have been employed for:

1. characterizing natural and human resources;
2. identifying policy and economic signals that affect the use of resources; analysing the responses of different socio-economic groups to these signals; and the environmental and development impacts of such responses;
3. defining priority environment and development issues;
4. developing optional solution packages to move towards sustainability – covering policy, planning, institutional, technical and social changes;
5. consensus-building to select and reconcile solution packages in a coherent strategy for sustainable development;
6. official and public approval and implementation;
7. review of development and conservation progress.

This process (see IUCN/WWF/UNEP, 1991) has been employed in several countries. It appears to be most successful where there is: high-level political backing from the beginning of the exercise; a strong ethic for cross-sectoral/institutional collaboration; extensive public consultation, treated as seriously as professional consultation; and effective structures for consensus-building.

Such prerequisites are often obtained in small islands. Smallness permits relatively easy liaison with island colleagues freely, consultation with the public, consensus-building, and rapid societal changes as a result of consensus. There are certain constraints, however. In small societies, intense face-to-face personalism and kinship ties can reduce objectivity in decision-making and inhibit the confrontation of serious (polarizing) issues (Benedict, 1967, quoted in McElroy *et al.*, 1987). The NCS framework encourages objectivity, but it remains to be seen whether it will be enough to encourage the confrontation of serious issues. Nevertheless, the NCS process – further developed to encompass sustainable development, and notably economic and social issues – would appear to provide a promising model for developing sustainable development strategies for small islands.

10.4.2 Developing the analytical tools

Effective analytical tools and planning capabilities may not be available readily in small islands. And, even if they are available, such tools as environmental impact assessment (EIA), cross-sectoral policy analysis, specific environmental economics techniques, landscape/seascape analysis and carrying capacity assessment, and the means to use these tools interactively, such as geographic information systems (GIS), tend to have been developed for larger countries. There is, however, a small, but useful, body of literature providing guidelines for environmental impact analyses in islands. (See e.g. IUCN, 1974, Caribbean Conservation Association, 1984, Office of Technology Assessment, 1987, and Beller *et al*, 1990.)

Nevertheless, many tools may need to be specifically adapted for use in islands, drawing from the disciplines of island biogeography, island and coastal zone ecology, resource and environmental economics, and island development studies, in a way that will make them practicable for necessarily limited island planning capabilities. Existing tools would be more effective if a strong analytical framework for examining island behaviour is developed; this has yet to emerge (McElroy *et al*, 1987). It could usefully be based on agroecosystems analysis participatory enquiry and environmental economics, applied to all island subsystems (see Chapter 4 for a discussion of agroecosystems analysis). There is hence a challenging research task. It cannot be done by individual small islands in isolation. This provides further justification for reducing the isolation of professionals on different islands through networking, professional associations and joint facilities.

10.4.3 Island resource management

Knowledge of resource management specific to small tropical islands is weak. Certain principles can be recommended, however.

From an ecological viewpoint, resource management should mimic and/or act in concert with natural energy buffers and ecological processes, especially to increase resilience in the face of external economic and environmental forces. Management systems would therefore include: constant ground cover; diversity of species and plant canopy architecture – such as in agroforestry; maximum use of solar energy and minimal use of external energy sources; ecological loop-closing (e.g. reusing wastes, transfer of organic matter and minerals between agroecosystems rather than loss to the sea). Traditional techniques may provide insight for such systems, although those that depend upon time, e.g. fallowing systems, may not be appropriate in densely-populated islands subject to high resource demands.

From an economic viewpoint, resource management should aim to sustain high income on relatively small land units. It should realize comparative advantages to produce speciality crops for which there are possibilities of creating monopolies, such as flower/fruit/vegetable/spice intensive horticulture and orchid cultivation. It should also realize comparative advantages for speciality services, such as ecotourism, marine sports and marine navigation. To do so will require good market information and promotion, and ensuring adequate and regular supplies – which, however, would tend to exclude commercial activities on the smallest islands. Especially in islands involved in tourism, a landscape management approach to land-based resources, based on local traditions, should add value.

From a social viewpoint, resource management should: maximize the potential reversibility of the land, e.g. putting as little fertile land as possible under concrete; providing for essential food security as well as for export – again, favouring multiple use; allowing resources to be left unmanaged for long periods while opportunities for alternative income-earning arise – favouring perennial crops; and sustaining cultural traditions (traditional landscape, cuisine, etc.).

In light of these principles, research is required for improving management techniques and yield regulation guidelines – especially for reef management, offshore fisheries and all multiple use regimes; and designing mitigation strategies for sea level rise and climate change.

REFERENCES

Bass, S.M.J. and Cambers, G. (1991) *Sea Level Rise and Climate Change in the Organisation of Eastern Caribbean States*, Phase I Report, Overseas Development Administration, London.

Beller, W., d'Ayala, P. and Hein, P. (eds) (1990) *Sustainable Development and Environmental Management of Small Islands*, Man and the Biosphere Series Vol. 5, Unesco, Paris.

Benedict, B. (1967) *Problems of Smaller Territories*, Athlone, London.

Brookfield, H.C. (1986) An Approach to Islands, paper presented at the Interoceanic Workshop on Sustainable Development and Environmental Management of Small Islands, Puerto Rico 3–7 Nov.

Brown, S. (1982) *An Overview of Caribbean Forests* (eds. A.E.S. Brown and A.E. Lugano) Proceedings of the First Workshop of Caribbean Foresters.

Caribbean Conservation Association (1984) *Environmental Guidelines for Development in the Lesser Antilles*, Caribbean Environment Technical Report No.3.

Crosby, A.W. (1986) *Ecological Imperialism: the Biological Expansion of Europe 900–1900*, Cambridge University Press, Cambridge.

Dahl, A.L. (1984) Oceania's Most Pressing Environmental Concerns, *Ambio* **13** (5–6).

Daly, H.E. (1984) *The Steady-State Economy: Alternative to Growthmania*, paper presented at The Other Economic Summit, London

Doumenge, F. (1985) The Viability of Small Intertropical Islands, in *States, Microstates and Islands*, (eds E. Dommen and P. Hein) Croom Helm, London.

Economist, The (1988) A Survey of the Caribbean, 6th August.

Girvan, N.P. (1991) Economics and the Environment in the Caribbean: an Overview, in *Caribbean Ecology and Economics*, (eds. N.P. Girvan and D. Simmons), Institute of Social and Economic Research, Kingston.

Grove, R. (1990) The Origins of Environmentalism *Nature* **345,** 11–14.

Hamnett, M.P. (1986) Pacific Islands Resource Development and Environmental Management, paper presented at the Interoceanic Workshop on Sustainable Development and Environmental Management of Small Islands, Puerto Rico 3–7 Nov.

Hein, P.L. (1986) Between Aldabra and Nauru, paper presented at the Interoceanic Workshop on Sustainable Development and Environmental Management of Small Islands, Puerto Rico 3–7 Nov.

Ince, M. (1990) *The Rising Seas*, Earthscan, London.

IUCN (1974) *Ecological Guidelines for Island Development*, Morges, Switzerland.

IUCN/WWF/UNEP (1991) *Caring for the Earth: a Strategy for Sustainable Living*, Gland, Switzerland.

Knight, F.W. and Palmer, C.A. (eds) (1989) *The Modern Caribbean*, Chapel Hill, University of North Carolina Press.

Liew, J. (1986) Sustainable Development and Environmental Management of Atolls, paper presented at the Interoceanic Workshop on Sustainable Development and Environmental Management of Small Islands, Puerto Rico 3–7 Nov.

McElroy, J. and de Albuquerque, K. (1991) Tourism Styles and Policy Responses in the Open Economy–Closed Environment Context, in *Caribbean Ecology and Economics*, (eds N.P. Girvan and D. Simmons), Institute of Social and Economic Research, Kingston.

McElroy, J., de Albuquerque, K. and Towle, E.L. (1987) *Old Problems and New Directions for Planning Sustainable Development in-Small Islands*, Ekistics 323/324

McElroy, J., Potter, B. and Towle, E.L. (1986) Challenges for Sustainable Development in Small Caribbean Islands, paper presented at Interoceanic Workshop on Sustainable Development and Environmental Management of Small Islands, Puerto Rico, 3–7 Nov.

Office of Technology Assessment, Congress of the United States (1987) Integrated Renewable Resource Management for US Insular Areas.

Ogden, M.R. (1989) The Paradox of Pacific Development, *Development Policy Review* 7 361–73.

Organization of American States (OAS) Executive Secretariat for Economic and Social Affairs (1988) Carriacou Integrated Development Programme, Grenada: Department of Regional Development Integrated Development Project.

Vermeiren, J.C. (1991) Natural Disasters: Linking Economics and the Environment with a Vengeance, in *Caribbean Ecology and Economics* (eds N.P. Girvan, and D. Simmons), Institute of Social and Economic Research, Kingston.

Watts, D. (1987) *The West Indies: Patterns of Development, Culture and Environmental Change since 1492*, Cambridge University Press, Cambridge.

Weiskel, T.C. (1990) *The Anthropology of Environmental Decline: Historical Aspects of Anthropogenic Ecological Degradation*, Reference Services Review, USA, Summer 1990.

Wood, D. (1968) *Trinidad in Transition: the Years after Slavery*, Oxford University Press, Oxford.

11

Sustainable economic development: economic and ethical principles

R.K. Turner and D.W. Pearce

11.1 INTRODUCTION

The current concern with global environmental issues – climate change, biodiversity loss, ozone layer depletion, etc. – reflects the evolution in thinking about environmentalism which has taken place over the last twenty years or so. During the 1970s concern was overtly focused on source limits, i.e. population growth and natural resources and food supply, with relatively less emphasis on sink limits, i.e. pollution and the assimilative capacity of the biosphere. By the time the United Nations Conference on Environment and Development (UNCED) had taken place in the summer of 1992, the primary focus for concern had shifted towards sink limits. Hence UNCED concentrated on two main issues, for which international agreements were signed, climate change and biodiversity.

Thus the 'Limits to Growth' debate has been superseded by the 'Global Environmental Change' debate. Nevertheless, the various competing world views, underlying ethics and related policy prescriptions, which the debates have highlighted or spawned, can still be broadly categorized as neo-Malthusian and neo-Ricardian perspectives, see Figure 11.1.

Just as the neo-Malthusian source limits argument, in the 1970s, stimulated a critique based on neo-Ricardian ideas linked to the effectiveness of scarcity signals (prices) and mitigation mechanisms (technical change and substitution); so the sink limits argument of the 1990s is being countered by technocentrist thinking. From this latter perspective, the extent and severity of the global environmental constraints

Economics and Ecology: New frontiers and sustainable development.
Edited by Edward B. Barbier. Published in 1993 by Chapman & Hall, 2–6 Boundary Row, London SE1 8HN. ISBN 0 412 48180 4.

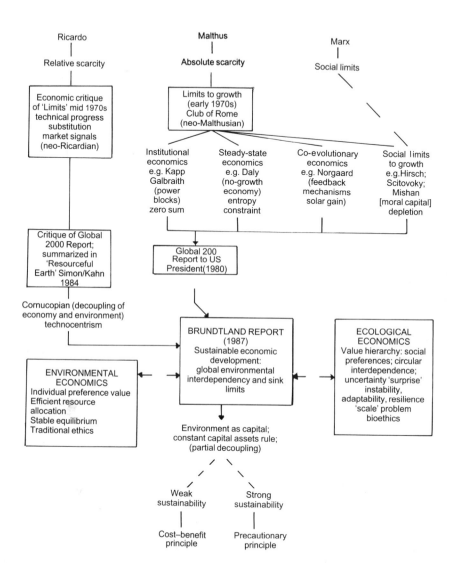

Figure 11.1 Development of environmental ecological economics. Partial and truncated chronology.

on economic growth have yet to be accurately quantified. Further, in the sub-context of raw materials supply, 'scarcity' is technocentrically a somewhat ephemeral concept (Scott and Pearse, 1992). The economy's need for raw material inputs will continue to be met via an expanding range of sources and the invention of new combinations of new resources. As Scott and Pearse (1992) put it, from this position, the only ultimate resource is innovation.

While all sides recognize that at a fundamental level economic systems are underpinned by ecological systems, there is considerable disagreement about the degree of decoupling of the economy and the environment that is, or will become, feasible. Limits supporters believe that the scale of economic activities may already be close to a maximum safe level in terms of the throughput of matter and energy and the resulting waste loading of the biospherical sinks (Daly and Cobb, 1989).

Limits critics point to evidence which they say indicates that the long term price of most primary commodities is declining. The supply base is continually being expanded as technical progress allows lower and lower grade resources to be economically exploited. Technical progress also opens up increasing opportunities for resource substitution, often with an extra bonus of decreasing resource use intensity. The only exception (i.e. uncertain prospects for a 'backstop technology') seems to be energy resources. From this perspective, then, science and technology will be the key to harmonizing (decoupling) the economy with the environment in our current search for what has become known as sustainable development.

11.2 WEAK AND STRONG SUSTAINABILITY PARADIGMS

Sustainable development (SD) has become a catch-all phrase for forms of economic development which stress the importance of environmental quality and the conservation of Nature's assets (World Commission on Environment and Development 1987; Pearce, Markandya and Barbier, 1989; Turner, 1988; Barbier, Markandya and Pearce, 1990; Pearce and Turner, 1990). Definitions of sustainable development abound (see the list given by Pearce *et al.*, 1989), but our main interest in the concept in this chapter lies in its catalytic role as a means of exploring the interface between environmental economics, human ecology and ethics – see Figure 11.1.

In itself, sustainable development is easily defined. Most people have some idea of what it means for a society to 'develop': almost certainly it would be a society in which real per capita incomes rise over time, but it would also be a society in which there are continuous improvements in knowledge and health as measured by, say, literacy and life expectancy. Human freedoms would also surely be part of the meaning of development. A high income country with no democracy would not qualify as a developed nation. Sustainable development is then simply development that is sustained through time.

More interesting than defining sustainable development is determining the conditions for achieving it. We argue that the definition of sustainable development implies that the next generation should not be 'worse off' in development terms than this one. In turn, this means leaving the next generation with a stock of capital assets that provide them with the *capability* to generate at least as much development as is achieved by this generation. We focus on 'capital' because we believe this is the main constituent of human capacity, but we stress that, for us, as for many economists, capital refers not just to man-made capital (K_m), but also to the stock of knowledge and skills ('human capital', K_h) and to the stock of environmental capital or 'natural capital' (K_n). Expressed in these terms, sustainability becomes a matter of conserving the overall capital stock, $K_m + K_h + K_n$.

We do not dwell on technology or population growth. As far as technology is concerned any advance can of course mean that the next generation's capital will have higher productivity than this generation's. Technically, then, future capital stocks could be smaller but as capable of generating as much development as today. We acknowledge this potential and note that (a) technology will tend to be embodied in new capital, (b) the aim of sustainable development is to raise capital productivity, (c) not all technology is benign (think of CFCs), and (d) such increases in capital productivity will be necessary anyway to offset the negative consequences of population growth on sustainability.

Now a policy of conserving the overall capital stock is consistent with running down any part of it. It would, for example, be justified on this rule to run down the environment provided the proceeds of environmental degradation were reinvested in other forms of capital. The cost–benefit rule cautions against running down environmental capital unless the benefits of so doing outweigh the costs. By and large, therefore, the cost–benefit rules and this broad concept of sustainability (which we call 'weak sustainability') are consistent. The cost–benefit rule says that it is acceptable to have environmental degradation if the costs of that degradation are less than the benefits. The weak sustainability rule says that it is acceptable to run down environmental capital if some other form of capital is built up instead. But the overall value of the capital stock would be smaller if the costs of lost environmental capital are not more than compensated for by the benefits of building up other forms of capital. The two rules are consistent.

Most discussion of sustainable development assumes that it is not acceptable to run down environmental assets. The reasons for protecting natural capital are interesting because, we argue, they are the self-same reasons that justify a precautionary approach in many instances. The reasons are:

1. *Uncertainty*: if we do not know the consequences for human well-being of running down natural capital – because of our ignorance of how complex ecosystems work – then acting as if there was certainty could have major detrimental consequences. Uncertainty dictates caution.
2. *Irreversibility*: some of the consequences of our actions are irreversible, as with, say, species extinction or raising the earth's temperature. Both reduce future generations' choice about the kind of society they wish to live in. On grounds of intergenerational fairness (at least), irreversibility also dictates caution. Note that only natural capital has the attribute of irreversibility. Man-made capital can be increased or decreased at will.
3. *Life-support*: some ecological assets ('critical' K_n) serve life support functions. Removing them in a context where there is no man-made substitute means possible major harm for mankind. This is the non-substitutability argument.
4. *Loss aversion*: there is some strong evidence in economics and psychology that people are highly averse to environmental losses, i.e. they feel a natural right to their existing endowment of natural assets.

Simplifying matters somewhat, in this chapter we distinguish three basic world views[1] – the conventional economic paradigm, which we take to be illustrated by standard cost–benefit analysis in which efforts are made to compare gains and losses in utilitarian terms. A subsidiary requirement of cost–benefit procedures is that any given target for policy should be achieved at minimum cost. In turn, the minimum cost requirement favours the use of certain economic instruments which act primarily on prices in the economy; the sustainability paradigm (which has a 'weak' and 'strong' version) in which the utilitarian cost–benefit paradigm is first modified to allow for a concept of intergenerational equity. A constant capital assets rule ('weak' version) and the notion of critical natural capital assets, which are non-substitutable and therefore must be conserved ('strong' version) may also be introduced as constraints on cost–benefit analysis and the functioning of the economic system. The constant capital assets rule reflects a moral imperative to care for the next generation, and this imperative is not readily interpreted in terms of utilitarian gains and losses. It may well be the case that current generations will have to bear higher costs than future gains in order to maintain constant capital assets. The utilitarian rule would reject this 'trade' because future benefits are less than current sacrifice (Beckerman, 1992). The strong sustainability paradigm, thus allows for nonutilitarian values but remain anthropocentric. The 'fairness' in question is fairness between people; and the bioethics (very strong sustainability) paradigm, which seeks, in one version, to support a steady-state (minimum resource take) economy, as well as to encompass non-instrumental values in nature and consequently notions of rights and

---

[1] In practice there is a spectrum of perspectives which overlap each other.

interests for non-human components of the biosphere (Naess, 1973). The economist's difficulty with the spectrum of non-instrumental values (particularly those of a nonanthropocentric kind) is that due allowance for them in practical decision-making produces stultifying rules of behaviour. We would wish to add an additional argument that an adequate level of environmental protection can be achieved with recourse to noninstrumental value philosophy and that in any case such viewpoints carry with them debatable ethical implications at least as far as questions of social justice in developing countries are concerned.

In short, we argue for the strong sustainability paradigm as a means of integrating economic efficiency, intergenerational equity and the precautionary principle (because of the existence of 'critical' K_n). The resulting constant capital rule has two incidental effects:

1. It protects the environments of the poorest communities in the world who depend directly on those environments for fuel, water and food.
2. It protects the environments of sentient nonhumans and nonsentient things.

A vital sustainable economics principle is that natural resources and environments are multifunctional and represent significant sources of economic value. It is therefore vitally important that the environment is valued correctly and that these values are integrated into management policy.

11.3 THE VALUATION OF ENVIRONMENTAL RESOURCES

From the conventional economic perspective, the sustainability issue has at its core the phenomenon of market failure and its correction via 'proper' resource pricing. What is required is an intertemporally efficient allocation of environmental resources through price corrections based on individual preference value, e.g. Solow, 1974; Solow, 1986. A vast literature has therefore grown upon the various monetary valuation methods and techniques available to 'price' the range of environmental goods and services provided by the biosphere (i.e. market-adjusted, surrogate-market and simulated-market methods).

According to conventional economic theory, the value of all environmental assets is measured by the preferences of individuals for the conservation of these assets. Thus individuals have a number of held values which in turn result in objects being given various assigned values. In order to arrive at a measure of total economic value, economists begin by distinguishing user values from non-user values. In a straightforward sense, user values derive from the actual use of the environment. Slightly more complex are values expressed through options to use the environment (option values). They are essentially expressions of preference (willingness to pay) for the preservation of an environment against some probability that the individual will make use of it at a later date. Provided the uncertainty concerning future use is an uncertainty relating to the 'supply' of the environment, economic theory indicates that this option value is likely to be positive. A related form of value is bequest value, a willingness to pay to preserve the environment for the benefit of one's children and grand-children.

Non-use values present more problems. They suggest values which are in the real nature of the thing but unassociated with actual use, or even the option to use the thing. Instead such values are taken to be entities that reflect people's preferences, but include concern for, sympathy with, and respect for the rights or welfare of nonhuman beings. Individuals may value the very existence of certain species or whole ecosystems. Total economic value is then made up of actual use value plus option value plus existence value.

A certain amount of progress has been made by economists attempting to determine empirical (monetary) measures of both environmental use values and non-use values. None of the techniques that have been utilized are problem free but enough empirical work has been undertaken to indicate that humans do value the environment positively. While the estimates made so far are subject to quite wide error margins, no one can doubt that the values uncovered are real and important.

Valuation techniques, notably the contingent valuation (CVM) approach, have been more extensively used, and many would argue, have been substantially improved, during the 1980s. Recent debates, both within the economics profession and between economists and non-economists, about the 'usefulness' (reliability and validity) of CVM have brought to the surface several general and fundamental questions. The theory, methodology and application of (nonmarket) value measurement of environmental resources have all come under scrutiny. Harris and Brown (1992), for example, have noted that, who gains or loses from some environmental change and how feelings of responsibility and self-interest influence the value judgements of these gainers and losers, have important implications for CVM.

There is now a large body of research which elicits individuals' valuations of changes in some environmental resources using the CVM, in which individuals express their preferences by answering questions about hypothetical choices. CVM has been subject to criticism, particularly as a result of theoretical and experimental research by psychologists and economists into the problem of eliciting preferences. Supporters of CVM are currently attempting to address both reliability and validity questions.

A basic question for the implementation of the CVM is whether willingness to pay for benefits (WTP) is the most appropriate indicator of value in a given situation. For cost–benefit analysis based on the Hicks–Kaldor compensation test, WTP would seem to be the appropriate measure for gainers from some resource reallocation decision, and WTA the proper measure for losers from that same reallocation. But as Harris and Brown (1992) have pointed out, it is often not easy to identify conclusively gainers and losers since this judgement is itself influenced by the valuer's own perspective.

Willig (1976) claimed that WTP and WTA measures should, in the absence of strong income effects, produce estimates of monetary value that are fairly close (within 5%). However, since 1976 strong evidence has been accumulated which shows that for given environmental goals, WTA is significantly greater than WTP (40%+ divergence). In addition, WTA valuations seem to have greater variance than WTP ones, and are less accurate predictors of actual buying/selling decisions.

The format of the questions used to elicit valuations may be continuous (or 'open-ended'), i.e. asking respondents to state WTP or WTA without any prompts concerning possible answers, or *discrete* (or 'dichotomous'), i.e. presenting the respondent with a single buying price or selling price which must be accepted or rejected. Many intermediate formats are also possible, e.g. bidding games. These differences in format can produce systematically different responses (Desvousges *et al*, 1987; Loomis, 1990).

A number of explanations have been offered for the differences in valuations elicited by different formats:

1. There may be income effects, as predicted by Hicksian consumer theory. In a recent paper Hanemann (1991) has argued that such effects could account for some observed WTP/WTA differences for public goods. He has calculated that a WTA measure five times greater than WTP can be justified in cases where the elasticity of substitution is low and/or the WTP/income rates is high i.e. for unique, irreplaceable environmental assets which individuals care a great deal about.
2. A psychological phenomenon, loss aversion, may be important especially in the case of potential losers in a resource change when WTA questions are related to giving up things, rights or privilege. Valuations may be made relative to reference points, losses being weighted more heavily than gains. Such effects, which could account for some WTP/WTA differences, have been found experimentally (e.g. Knetsch and Sinden, 1984). Similarly anchoring effects (or starting point bias) may cause differences between responses to discrete and continuous formats (Green and Tunstall, 1991).
3. WTA questions may be less readily understandable than WTP ones, since most people have more experience of buying goods, paying taxes, etc. than of selling. Similarly, continuous questions may be less readily understandable than discrete ones, since most people have more experience of choosing whether or not to pay stated prices than of stating valuations.

4. The continuous format may have a stronger tendency than the discrete format to suggest opportunities for free riding.
5. Respondents may act strategically, i.e. make guesses about how their answers will be used and then give the answers that they believe will serve their interests best.

Overall, it is likely that merely identifying gainers and losers in some resource change situation will be insufficient to determine whether WTP or WTA is the most appropriate indicator of value. We need to know more about the motives of the valuer (Harris and Brown, 1992). Economics has much to learn from psychological research in this context. In fact, some of CVM's strongest critics are to be found outside the economics profession, in the ranks of the philosophers, psychologists, political scientists and scientists.

11.3.1 Debate outside economics

Sagoff (1988) has argued that economics makes a 'category mistake' in its approach to environmental valuation. For him, it is not preferences but attitudes that determine people's environmental valuations. Thus people may not be willing to consider market-like transactions (assumed by CVM) involving public resources. CVM surveys pick this effect up in the form of refusals and 'protest bids'. Some combination of individual preferences and public (collectively held) preferences will be held by any given individual who by necessity has to operate in daily life as both a consumer and a citizen. Thus the environment can be both a purchased commodity and a moral/ethical concern.

According to Sagoff environmental economics has no role to play in the determination of the goals of environmental policy. Environmental protection standards are determined by political, cultural and historical factors not by preference-based values. If economics has a role it is restricted to revealing the costs (social opportunity costs) of the pre-emptive environmental standards. But if action is taken on the basis of the opportunity cost analysis then an implicit valuation has been made. Nevertheless, from this viewpoint, there is no role for direct monetary valuation (preference-based) of the benefits of environmental protection policy.

Other critics do not go as far as completely rejecting the validity of WTP/WTA measures of value, but instead argue that economic values are only partial values for many environmental resources. Thus Brennan (1992) seeks to distinguish so-called transformative value from the economic value of environmental assets.

Many environmental assets, according to Brennan, possess additional transformative properties such that 'exposure' to the assets causes a change in people's preferences. The impact of transformative values on people's preferences is said to be completely unpredictable and the degree of impact (when it occurs) will vary significantly from person to person. If natural things and systems possess transformative values, then it is argued they cannot be priced by economic analysis.

As far as we can see this transformative value argument can be accommodated within the total economic value approach. The transformative value is equivalent to a use or existence value, only it is latent, and for some individuals may never actually exist, i.e. their preferences are never changed (transformed) by contact with or knowledge of the specific thing that possesses the transformative property. At the policy level, if it is the case that natural things and systems possess transformative values then a conservation strategy based on the total economic value principle (in particular option and bequest values) would be sufficient to guarantee their future existence. So individuals may value (exhibit a WTP for) natural things and systems in order to retain the option to use them, or be transformed by them, some time in the future. Bequest value similarly expresses an individual's wish to retain options for their descendants.

Some scientists have argued that the full contribution of component species and processes to the aggregate life-support service provided by ecosystems has not been captured in economic values (Ehrlich and Ehrlich, 1992). There does seem to be a sense in which this scientific critique of the partial nature of economic valuation has some validity; not in relation to individual species and processes but in terms of the prior value of the aggregate ecosystem structure and its life-support capacity.

Since it is the case that the component parts of a system are contingent on the existence and functioning of the whole, then putting an aggregate value on ecosystems is rather more complicated a matter than has previously been supposed in the economics literature. Taking wetland ecosystems as an example, the total wetland is the source of **primary value** (PV) (Turner, 1992). The existence of the wetland structure (all its components, their interrelationships and the interrelationships with the abiotic environment) is prior to the range of function/service values. The concept of total economic value (TEV) has two not just one limitation as previously supposed. TEV may fail to fully encapsulate the total secondary value (TSV) provided by an ecosystem, because in practice some of the functions and processes are difficult to analyse (scientifically) as well as to value in monetary terms. But in addition, TEV fails to capture the PV of ecosystems, indeed this or 'glue' value notion is very difficult to measure in direct value terms since it is a non-preference, but still instrumental, type of value.

We believe that this primary and secondary ecosystem value classification goes some way towards satisfying many scientists' concerns about the 'partial' nature of the conventional economic valuation approach (Ehrlich and Ehrlich, 1992). It is also a classification that avoids the instrumental versus non-instrumental value in nature debate which we believe has become rather sterile.

More formally:

> each ecosystem provides a source or stock of primary value = total TPV = 'glue' value of the ecosystem;
> the existence of a 'healthy' ecosystem (i.e. one which is stable and/or resilient) provides a range of functions and services (secondary values) = TSV;
> Total Ecosystem Value = TV;
> and TV = TVP + TSV;
> TSV = TEV (total economic value)
> where TEV = UV (use value) + NUV (non-use value);

What is clear is that the components of TEV (use and non-use values) cannot simply be aggregated. There are often trade-offs between different types of use value and between direct and indirect use values. Smith (1992) has also pointed out that the partitioning of use and non-use values may be problematic, if it is the case that use values may well depend on the level of services attributed to non-use values. The TEV approach has to be used with care and with a full awareness of its limitations.

11.3.2 Towards an understanding of the valuation process

Figure 11.2 summarizes the various elements thought to comprise the full valuation process. The interaction between a person and an object (to be valued) involves perception of the objects and a process whereby relevant held values, beliefs and dispositions come to the forefront. Perception and beliefs are interrelated and together result in an unobservable sense of value (utility), which may then be expressed as an assigned value and certain behaviour (Brown and Slovic, 1988). Brown and Slovic conclude that the valuation context may affect how objects are perceived, the beliefs that become relevant, the utility experienced and the value assigned.

Information (existing and new) plays a key role in the valuation process. An individual's familiarity with the environmental commodity/context and the resulting perceptions are dependent on both the information stock and the provision of new information. The type and form of information supplied is particularly important in situations where direct perception is not possible and recourse to 'expert' knowledge is required.

Perceptions, information and beliefs all then feed into motivation. Harris and Brown (1992) identify what they call a responsibility motive in the environmental loss context. The motive is best represented as a spectrum of feelings extending from personal responsibility to a more general concern for the environment unrelated to use value. Randall (1987) has argued that all non-use values have their basis in the motive of altruism –

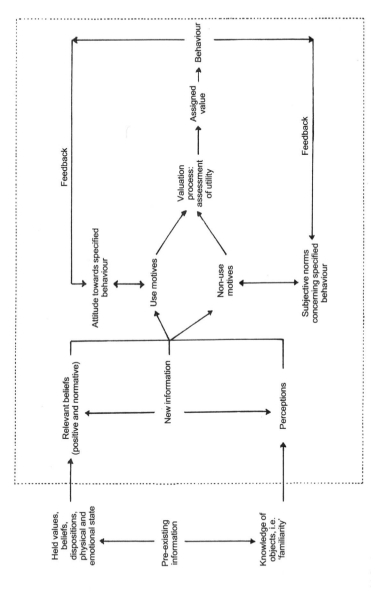

Figure 11.2 An expanded framework for assignment of value to objects and specified behaviour. Adapted from Harris and Brown (1992) and Ajzen and Fishbein (1977).

interpersonal, intergenerational and Q-altruism (based on the knowledge that some asset Q itself benefits from being undisturbed). What this discussion of motivation does is to question the simplistic 'rational economic person' psychological assumptions that underpin conventional economic analysis. The motive of self-interest is only one of a number of human motivations and need not be the dominant one.

Maslovian psychology, for example, substitutes the concept of human needs for human wants and portrays needs in a hierarchical structure (Maslow, 1970) see Figure 11.3. Instead of an individual facing a flat plane of substitutable wants (as in conventional economics), Maslow conceives of the same individual attempting to satisfy levels of need. The satisfaction of higher level needs leads to a process of 'self-actualization'. Self-actualized individuals would be expected to possess a strong responsibility motivation and hold non-use values. Such individuals might well be prepared to pay to maintain some environmental asset regardless of the benefits they themselves receive from that asset.

Because of the divergence between WTP and WTA valuations many practitioners have taken the pragmatic decision to regard stated WTP valuations as reliable measures of true WTP and therefore to use CVM only in cases in which WTP is the appropriate measure of benefit. But then the

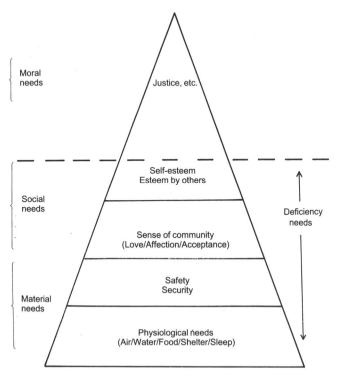

Figure 11.3 Maslow's hierarchy of needs (values). Source: Maslow (1970)

question becomes what is the exact set of cases in which WTP is appropriate? Harris and Brown (1992) have argued that WTP is in fact the appropriate measure of welfare change for a majority of situations. They identify only self-interested losers from a resource change as the appropriate group to be surveyed with WTA format questions. A mail survey undertaken (in 1988) by the same researchers indicated that 53% of the taxpayers' sample said that their state of Idaho should pay for the loss of nongame wildlife with tax dollars, implying that all taxpayers should pay to cover this loss (a WTP rather than WTA approach). Only 32% of survey respondents said that only those responsible for the loss should pay to prevent it (WTA format). Thus altruism and moral responsibility may well have an important role to play in influencing policy judgements.

11.4 SUSTAINABILITY AND ETHICS

SD is future oriented in that it seeks to ensure that future generations are at least as well off, on a welfare basis, as current generations; it is therefore in economic terms a matter of intergenerational equity and not just efficiency. The distribution of rights and assets across generations determines whether the efficient allocation of resources sustains welfare across human generations (Howarth and Norgaard, 1992). The ethical argument is that future generations have the right to expect an inheritance sufficient to allow them the capacity to generate for themselves a level of welfare no less than that enjoyed by the current generation. What is required then is some sort of intergenerational social contract.

SD also has a poverty focus, which in one sense is an extension of the intergenerational concern. Daly and Cobb (1989) have argued that families endure over intergenerational time. To the extent that any given individual is concerned about the welfare of his/her descendants, he/she should also be concerned about the welfare of all those in the present generation from whom the descendant will inherit. Accordingly, a concern for future generations should reinforce and not weaken the concern for current fairness. Ethical consistency demands (despite the trade-offs involved) that if future generations are to be left the means to secure equal or rising per capita welfare, the means to maintain and improve the well-being of today's poor must also be provided. Collective rather than individual action is required in order to effect these socially desirable intra- and intergenerational transfers. In any case, it seems to us, that no nation that neglects the most vulnerable in society ought to be labelled 'developed' or 'developing'.

The degree of concern, as expressed by the rate of time discount (DR) attached to the welfare of future generations, that is ethically required of the current generation is another controversial matter. Six positions seem possible – moral obligations to the future exist, but the welfare of the future is less important than present welfare ($0 < DR < \infty$); moral obligations to the future exist and the future's welfare is almost as important as present welfare (social time preference rate = STPDR; STPDR > 0 < DR); a discounting procedure is only acceptable after the imposition of pre-emptive constraints on some forms of economic development; obligations to the future exist and the future is assigned more weight than the present (DR is negative); rights and interests of future people are exactly the same as contemporary people (DR = 0); there is no obligation at all on the present to care about the future (DR = ∞).

For many commentators traditional ethical reasoning is faced with a number of challenges in the context of the sustainable development debate. Ecological economists would argue that the systems perspective demands an approach that privileges the requirements of the system above those of the individual. This will involve ethical judgements about the role and rights of present individual humans as against the system's survival and therefore the welfare of future generations. We have also already argued that the poverty focus of sustainability highlights the issue of inter-generational fairness and equity.

So 'concern for others' is an important issue in the debate. Given that individuals are, to a greater or lesser extent, self-interested and greedy, sustainability analysts explore the extent to which such behaviour could be modified and how to achieve the modification (Turner, 1988; Pearce, 1992). Some argue that a stewardship ethic (weak anthropocentrism, Norton 1987) is sufficient for sustainability, i.e. people should be less greedy because other people (including the world's poor and future generations) matter and greed imposes costs on these other people. Bioethicists would argue that people should also be less greedy because other living things matter and greed imposes costs on these other non-human species and things. This would be stewardship on behalf of the planet itself (Gaianism) in various forms up to 'deep ecology' (Naess, 1973; Turner, 1991).

The degree of intervention in the functioning of the economic system deemed necessary and sufficient for sustainable development also varies across the spectrum of viewpoints. Supporters of the steady-state economy (extensive intervention) would argue that at the core of the market system is the problem of 'corrosive self-interest'. Self-interest is seen as corroding the very moral context of community that is presupposed by the market. The market depends on a community that shares such values as honesty, freedom, initiative, thrift and other virtues whose authority is diminished by the positivistic individualistic philosophy of value (consumer sovereignty) of conventional economics. If all value derives only from the satisfaction of individual wants then there is nothing left over on the basis of which self-interested, individualistic want satisfaction can be restrained (Daly and Cobb, 1989).

Depletion of **moral capital** (K_e) may be more costly than the depletion of other components of the total capital stock (Hirsch, 1976). The market does not accumulate moral capital, it depletes it. Consequently, the market depends on the wide system (community) to regenerate K_e just as much as it depends on the ecosystem for K_n.

Individual wants (preferences) have to be distinguished from needs. For humanistic and institutional economists, individuals do not face choices over a flat plane of substitutable wants, but a hierarchy of needs. This hierarchy of needs reflects a hierarchy of values which cannot be completely reduced to a single dimension (Swaney, 1987). Sustainability imperatives therefore represent high order needs and values.

11.5 CONCLUSIONS

The issue of development versus conservation poses some of the most complex problems for modern environmental economics and ethics. In this chapter we have contrasted three broad paradigms for the decision-making process. The benefit–cost approach, rooted in utilitarianism, is an approach which evaluates the problem from the standpoint of economic efficiency. It does not, in its basic form anyway, consider other social objectives such as equity, duty or moral obligation. The cost–benefit approach is deliberately anthropocentric, but it does not exclude judgements on behalf of other species, the 'stewardship' motive. The phenomenon of 'existence value' identified and measured by environmental economists may well capture such values. The second paradigm identified was the modified cost–benefit approach which is derived from the concept of sustainable development.

Sustainability requires, we argue, a commitment to the conservation of capital stocks. In the weak sustainability variant cost–benefit analysis is modified through the idea of offsetting or compensating investments designed to compensate for cumulative damage done by investments which pass the orthodox cost–benefit test. In the strong sustainability variant the precautionary principle and the notion of critical natural capital assets provide further constraints on economic policy. Our interest in the constant capital rule lies in its role as a means of securing nonutilitarian values, including intergenerational equity, concern for the disadvantaged in current society, sentient nonhumans and nonsentient things. From the weak sustainability position the protective nature of the rule is an incidental effect of its primary purpose – economic efficiency and intergenerational equity.

Thirdly, the bioethical standpoint argues either for some broad equality between anthropocentric values and 'intrinsic' values in nature, or for 'higher moral ground' for intrinsic values. Intrinsic values in this content are 'in' being and objects rather than 'of' human beings. We argue against the bioethical standpoint on three grounds. It is stultifying of development and therefore has high social costs in terms of development benefits forgone. It may be conducive to social injustice by defying development benefits to the poorest members of the community, now and in the future.

It is redundant in that the modified sustainability approach generates many of the benefits alleged to accrue from the concern for intrinsic (non-instrumental) values.

REFERENCES

Ajzen, I. and Fishbein, M. (1977) Attitude-behaviour relations: a theoretical analysis and review of empirical research. *Psychological Bulletin*, **84**, 888–918.

Barbier, E.B., Markandya, A. and Pearce, D.W. (1990) Environmental sustainability and cost-benefit analysis. *Environment and Planning*, **22**, 1269–76.

Beckerman, W. (1992) Economic growth and the environment: Whose growth? Whose environment? *World Development*, **20**, 481–96.

Brennan, A. (1992) Moral pluralism and the environment. *Environmental Values*, **1**, 15–33.

Brown, T.C. and Slovic, P. (1988) Effects of context on economic measures of value, in: *Amenity Resource Valuation: Integrating Economics with Other Disciplines*, (eds G.L. Peterson, B.L. Driver and R. Gregory) Venture, State College, PA, pp 23–30.

Daly, H.E. and Cobb, J.B. (1989) *For the Common Good: Redirecting the Economy Towards Community, the Environment and a Sustainable Future*, Beacon Press, Boston.

Desvousges, W.H. *et al.* (1987) Option price estimates for water quality improvement. *Journal of Environmental Economics and Management* **14**, 248–67.

Ehrlich, P.R. and Ehrlich, A. (1992) The value of biodiversity, *Ambio*, **21**, 219–26.

Green, C.H. and Tunstall, S.M. (1991) The evaluation of river quality improvements by the contingent valuation method. *Applied Economics*, **23**, 1135–46.

Hanemann, W.M. (1991) Willingness to pay and willingness to accept: How much can they differ? *American Economic Review*, **81**, 635–47.

Harris, C.C. and Brown, G. (1992) Gain, loss and personal responsibility: the role of motivation in resource valuation decision-making. *Ecological Economics*, **5**, 73–92.

Hirsch, F. (1976) *Social Limits to Growth*, Routledge, London.

Howarth, R.B. and Norgaard, R.B. (1992) Environmental valuation under sustainable development. *American Economic Review Papers and Proceedings*, **82**, 473–7.

Knetsch J. and Sinden J. (1984) Willingness to pay and compensation demanded: experimental evidence of an unexpected disparity in measures in value. *Quarterly Journal of Economics*, **99**, 507–21.

Loomis, J.B. (1990) Comparative Reliability of the Dichotomous Choice and Open-Ended Contingent Valuation Technique. *Journal of Environmental Economics and Management*, **18**, 78–85.

Maslow, A. (1970) *Motivation and Personality*, Harper and Row, New York.

Naess, A. (1973) The shallow and the deep, long range ecology movement: a summary. *Inquiry* **16**, 95–100.

Norton, B.G. (1987) *Why Preserve Natural Variety?* Princeton University Press, Princeton NJ.

Pearce, D.W. (1992) Green economics. *Environmental Values*, **1**, 3–13.

Pearce, D.W. Markandya, A. and Barbier, E.B. (1989) *Blueprint for a Green Economy*, Earthscan, London.

Pearce, D.W., and Turner R.K. (1990) *Economics of Natural Resources and the Environment*, Harvester Wheatsheaf, Hemel Hempstead and London.

Randall, A. (1987) The total value dilemma in: *Toward the Measurement of Total Economic Value*, (eds G.L. Peterson and C.F. Sorg) USDA For. Serv. Gen. Tech.

Rep. Rm-148, Rocky Mountain Forest and Range Experiment Station, Fort Collins, CO, pp. 3–13.

Sagoff, M. (1988) Some problems with environmental economics. *Environmental Ethics*, **10**, 57–64.

Scott, A. and Pearse, P. (1992) Natural resources in a high-tech economy: scarcity versus resourcefulness. *Resources Policy*, **18**, 154–66.

Simon, J. and Kahn, H. (1984) *Resourceful Earth*, Blackwell, Oxford.

Smith., V.K. (1992) Non market valuation of environmental resources: An interpretative appraisal, draft copy of unpublished paper.

Solow, R.M. (1974) Intergenerational equity and exhaustible resources. *Review of Economic Studies Symposium*, 29–46.

Solow, R.M. (1986) On the intertemporal allocation of natural resources. *Scandinavian Journal of Economics*, **88**, 141–9.

Swaney, J. (1987) Elements of a neo-institutional environmental economics. *Journal of Environmental Issues*, **21**, 1739–79.

Turner, R.K. (1988) Wetland conservation: economics and ethics, in *Economics, Growth and Sustainable Environments*, (eds D. Collard *et al.*), Macmillan, London.

Turner, R.K. (1991) Economics and Wetland Management. *Ambio* **20**, 59–63.

Turner, R.K. (1992) Speculations on weak and strong sustainability, CSERGE GEC Working Paper 92–26, CSERGE, UEA, Norwich and UCL, London.

Willig, R.D. (1976) Consumer's Surplus without Apology. *American Economic Review*, **66**, 587–97.

World Commission on Environment and Development (1987) *Our Common Future*, Oxford University Press, Oxford.

12

Postscript

Edward B. Barbier

As discussed in the Introduction, the main purpose of this volume is to demonstrate the convergence of economic and ecological perspectives in furthering our understanding of what 'environmentally sustainable' development entails and how best to 'operationalize' it. The previous chapters have illustrated this theme through examining a variety of approaches and topics. Many important conclusions have been reached, which we hope are also relevant to a wider range of natural resource management problems other than those discussed here. Rather than repeat the main conclusions of each individual chapter, I would prefer to close this volume by summarizing some of the major themes emphasized by many, if not all, of the chapters.

First, the chapters in this volume have stressed the need to view many natural resource management problems in the context of an integrated economic–environmental system. The type of natural resource problems we are dealing with are not standard stock depletion or pollution problems, but tend to take the form of 'pervasive' environmental degradation and ecological disturbances that can push an economic–environmental system to its tolerance levels – and beyond them. This has implications for both economics and ecology, as in confronting these types of problems, both disciplines have had to recognize fundamental ecological–economic interdependencies in the parameters influencing the system.[1]

However, once the ecologist or economist has widened the scope of analysis to consideration of an integrated economic–environmental system,

[1] For the interested reader, I have written in more detail about the implications of this 'new' type of natural resource management problem for environmental and resource economics in Barbier, 1989 and 1990.

Economics and Ecology: New frontiers and sustainable development.
Edited by Edward B. Barbier. Published in 1993 by Chapman & Hall, 2–6 Boundary Row, London SE1 8HN. ISBN 0 412 48180 4.

then one is immediately confronted by the uncertainties surrounding eco-logical–economic interdependencies. These uncertainties are much more comprehensive than uncertainty over rainfall patterns and duration, crop yields, prices, and so forth. Instead, the key uncertainties seem to be about the extent to which ecological variables affect economic variables, and vice versa, so that we can determine the economic and ecological 'limits' that may bound the system. Failure to respect these limits may lead to the design of natural resource management policies and the encouragement of prac-tices that also push the system beyond 'sustainable' thresholds of tolerance.

Many authors have also indicated that sustaining the economic–environment system over the long run is only one of many possible social objectives for that system. Thus there are always inherent trade-offs be-tween sustainability and other system goals. The idea that species or specific resources may not be retained forever, i.e. that from society's perspective extinction may in some cases be 'optimal', is a familiar result in the economics of renewable resources.[2] In ecology, too, it has been recognized that the 'resilience' of an ecosystem is only one property of many, and there are inherent trade-offs among the various properties characterizing the system.[3] Ensuring the sustainability of a given eco-nomic–environmental system is again only one of many possible outcomes. What may determine whether this outcome is 'optimal' compared to others is just as much governed by ethical considerations, such as concern over the intergenerational distribution of income and over the 'total value' of natural environments and species, as by technical factors, such as the ability to substitute man-made for natural capital and the precise magnitude of 'ecological' carrying capacities.

Decisions to maintain the natural resource base supporting the system will therefore depend on the costs and benefits of depletion versus conser-vation. These are not easy to determine, given the uncertainties over ecological–economic interdependencies, the non-marketed nature of many environmental values and the pervasiveness of economic incentives that work against efficient and sustainable natural resource management. The private costs and benefits of environmental degradation may also be mis-leading because they tend to diverge significantly from the social costs and benefits. Thus, from a social perspective, it is imperative that decisions affecting natural resource management are guided by proper economic and ecological analysis of environmental values.

In short, because much excessive degradation of the environment results from individuals in the marketplace and from governments not fully recognizing and integrating environmental values into decision-making, it becomes even more essential that this problem is corrected. The result of

[2] For further discussion of this result, see Clark (1976); Dasgupta (1982); and Smith (1977). For applications to the case of elephants and ivory poaching, see Barbier *et al.*, (1990).

[3] See, for example, Holling (1973); Conway (1985) and Conway's Chapter 4 in this volume.

market and policy failures is a distortion in economic incentives. That is, the private costs of actions leading to environmental degradation do not reflect the full social costs of degradation, in terms of the environmental values forgone. There are several reasons for this outcome.

First, the market mechanisms determining the 'prices' for natural resources and products derived from conversion of natural resource systems do not automatically take into account wider environmental costs, such as disruptions to ecological functions, assimilative capacity, amenity values and other environmental impacts or forgone option and existence values – i.e. the value of preserving certain natural environments, species and resources today as an 'option' for future use or simply because their 'existence' is valued. Nor do market mechanisms account for any user cost – the cost of forgoing future direct or indirect use benefits from resource depletion or degradation today.

In addition, even the direct costs of harvesting resources or converting natural resource systems are often subsidized and/or distorted by public policies. As a result, individuals do not face even the full private costs of their own actions that degrade the environment. Unnecessary and excessive degradation ensues.

Viewing natural resource management problems in the context of economic–environment systems, recognizing and respecting the uncertainties surrounding ecological–economic interactions, exploring the welfare implications of sustainability and its potential trade-offs with other system goals, assessing and valuing properly environmental resources and functions, and finally, determining the implications of market and policy failures for the incentives for natural resource management are all important steps in the process of designing proper policies and investment strategies to control excessive environmental degradation. What we hope to have demonstrated in this volume is that both economics and ecology can make important contributions to this process – which we generally term 'environmentally sustainable' development – and more importantly, that the contribution is further enhanced if the two disciplines learn from and collaborate with each other. We hope that others will agree with us, and perhaps also be inspired to make similar contributions.

REFERENCES

Barbier, E.B. (1989) *Economics, Natural-Resource Scarcity and Development: Conventional and Alternative Views*, Earthscan, London.

Barbier, E.B. (1990) Alternative approaches to economic–environmental interactions, *Ecological Economics*, **2**, 7–26.

Barbier, E.B., Burgess, J.C., Swanson, T.M. and Pearce, D.W. (1990) *Elephants, Economics and Ivory*, Earthscan Publications, London.

Clark, C. (1976) *Mathematical Bioeconomics: The Optimal Management of Renewable Resources*, John Wiley, New York.

Conway, G.R. (1985) Agroecosystem analysis, *Agricultural Administration*, **20,** 31–55.
Dasgupta, P.S. (1982) *The Control of Resources*, Basil Blackwell, Oxford.
Holling, C.S. (1973) Resilience and the stability of ecological systems, *Annual Review of Ecological Systems*, **4,** 1–24.
Smith, V.L. (1977) Control theory applied to natural and environmental resources: An exposition, *Journal of Environmental Economics and Management*, **4,** 1–24.

Index